40p

AIR HISTOR[barcode]FORCE)

The history of Royal Air Force CRANWELL

Group Captain E B Haslam

NDON HER MAJESTY'S STATIONERY OFFICE

First published 1982

ISBN 0 11 772359 2

HER MAJESTY'S STATIONERY OFFICE

Government Bookshops

49 High Holborn, London WC1V 6HB
13a Castle Street, Edinburgh EH2 3AR
Brazennose Street, Manchester M60 8AS
Southey House, Wine Street, Bristol BS1 2BQ
258 Broad Street, Birmingham B1 2HE
80 Chichester Street, Belfast BT1 4JY

Government publications are also available through booksellers

Rehearsal for the Queen's Review, July 1981

Contents

List of Illustrations

Foreword

A Royal Naval Air Service station was established at Cranwell in 1915. Five years later one of Sir Hugh Trenchard's ambitions was realised when the Royal Air Force College for officer cadets was opened.

My family has been associated with developments at Royal Air Force Cranwell since the earliest days and it was at the Royal Air Force College that I received my pilot's wings in 1971.

The history of Cranwell forms a vital part of the history of the Royal Air Force itself and this book gives for the first time an account of developments at Cranwell from 1915 to the present day. As Patron of the Old Cranwellian Association I welcome it and hope it will give pleasure and revive memories both for former flight cadets of the Royal Air Force College and for all those others who served at Royal Air Force Cranwell.

Charles.

Acknowledgements

This book owes much to MRAF Sir Neil Cameron who was the Chief of the Air Staff and to Air Chief Marshal Sir Neville Stack who was the Air Secretary in 1977. They prepared the way for me to be invited officially to write the History of Royal Air Force Cranwell when I retired from the post of Head of the Air Historical Branch in 1978.

I have to acknowledge with thanks the gracious permission of Her Majesty the Queen to make use of material from the Royal Archives.

Every assistance has been given to me by the last three Commandants of the RAF College—Air Vice-Marshal W E Colahan, Air Vice-Marshal D Harcourt Smith and Air Vice Marshal B Brownlow.

Mrs Jean King ALA, the Librarian in College Hall, willingly made available papers, photographs and other resources in the College Library.

I was also privileged to study the minutes of meetings of the Old Cranwellian Association from 1926 onwards.

My basic source of information was the series of Operations Record Books (Forms 540) of all the units which had been formed or stationed at Royal Air Force Cranwell from 1915 to the present day. Many of the F540s are now in the Public Record Office at Kew but papers less than 30 years old are stored in the Ministry of Defence Archives at Hayes or held in the Air Historical Branch (RAF).

I am indebted to Air Commodore H A Probert and to his staff in AHB (RAF) for their ever willing support in response to my detailed enquiries. I am particularly grateful to Mr D C Bateman and also to Miss Joan Robinson of the General Staff Map Section, who helped to produce the illustrations and photographs.

The Viscount Trenchard allowed me to consult his father's private papers at Abdale House.

MRAF Sir Dermot Boyle, MRAF Sir Denis Spotswood and Air Chief Marshal Sir Neville Stack read the whole book in draft and made valuable suggestions and comments.

I am indebted to many Old Cranwellians and to retired officers and airmen who served at Cranwell. The following afforded me the privilege of oral interviews which were tape recorded: MRAF Sir Dermot Boyle, the late Group Captain F E Nuttall, Air Chief Marshal Sir Walter Merton, Air Chief Marshal Sir David Lee, Air Chief Marshal Sir Neville Stack, Air Chief Marshal Sir Lewis Hodges, the late Air Chief Marshal Sir John Stacey, Air Marshal Sir Edward Chilton, Air Vice-Marshal M D Lyne, Air Commodore E D Mac K Nelson, Air Commodore L G Levis, Group Captain G J H Jeffs, Wing Commander A J Payn and Squadron Leader R C L Parkhouse. There were literally hundreds more who would have helped but time was the enemy.

I also had interviews and talks with many civilians who had a knowledge of Cranwell in past years. Among these were: Lady George, Mrs Jill Lidderdale, Mrs Pam Smith, the late Miss Margaret Brackenbury, Miss Pat Halahan, Mrs Pauline Blackie (nee Adlercron), Captain Sir Anthony Thorold, RN (Retd), Mr A H Mather JP of Boston, Mr Stanley Robinson of Leasingham, Mr Cyril Hobbs of Penge, Mr J W Collishaw, Head Steward of College Hall Officers' Mess and Mr Ray Bush of the District Works Office, Cranwell.

Very helpful written contributions were sent in reply to my questions by Group Captain A H W J Cocks (1923 Entry), by Group Captain R C Travis who was serving at Cranwell in 1980 and by Mr L C Wigdor of Chislehurst.

Mr J C Andrews, Chief Librarian, at the Ministry of Defence, assisted my enquiries as did Mr J D Brown of the Naval Historical Branch, Mr H R Milton of the Adastral Library and the Librarian of the RAF Museum at Hendon.

All quotations from official documents not yet released to the Public Record Office are Crown Copyright and are reproduced by courtesy of The Controller, Her Majesty's Stationery Office.

Mr David James of Childe Okeford supplied me with photographs of a Cranwell Parade in 1920 and Miss Pat Halahan gave me photographs from her private album. Other photographs were prepared by the very efficient staff of CS (RepS) Pg.

I am grateful to Mrs Margaret Todd of Beckenham for her excellent typing and to Mr and Mrs J E Ashton of Wyberton, Boston for their hospitality during many research visits to the Lincolnshire area.

Introduction

If you travel by the route from the south which so many have used over the years, you turn off the Great North Road at Colsterworth. By the back roads you pass the grass airfield of Spitalgate and come to the village of Ancaster rich in Roman remains. Following the Ermine Street northwards for about three miles you top a rise and see in the distance a white dome surmounting an imposing building. This improbable sight amid the dappled and ordered Lincolnshire fields has evoked much stirring of emotion and memory since it rose in the early 1930s. C L M Brown wrote in Air Power in 1954:

> Coming from Newark as one surmounts the Jurassic Ridge at Caythorpe, the distant dome of the College catches the eye, brooding a little incongruously over the west Lincolnshire wastes.

From whichever direction you approach Cranwell the central pinnacle by day or the flashing beacon on top of the College by night takes the attention. But the history of the station or base at Cranwell is older than that of the College and this narrative seeks to tell the whole story of Cranwell which together with Halton has been one of the enduring centres of tradition in the Royal Air Force. The story begins in the autumn of 1915 and is still continuing. The College will feature prominently as the major unit of the station but there are other stories of other units which form part of the incongruous but successful union between the Service and the people of Lincolnshire who are proud of the camp that grew up near the village of Cranwell.

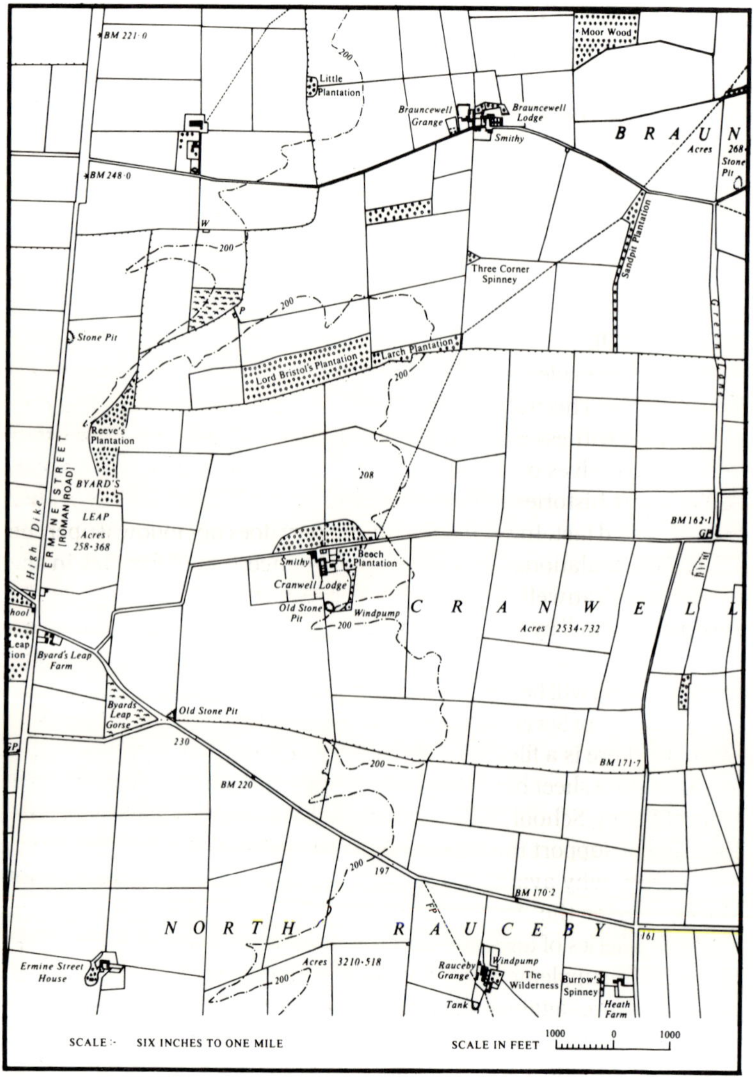

Fig 1 The land around Cranwell in 1916.

1 *Naval Beginnings— The Royal Naval Air Service 1915-1918*

Why Cranwell?

Documentary evidence on the early history of Cranwell is scarce and unsystematic. This narrative owes much in parts to the oral testimony of surviving witnesses and to the recollections of others whose written reports themselves owe more to memory than contemporary papers. Earlier short histories of Cranwell have been useful as guides rather than accepted fact. In particular this history does not follow Rupert de la Bère's speculations[1] on the mediaeval ancestors of families in the environs of Cranwell. Such history is better studied in books like *The Parish of Cranwell* by Harold Ison.[2] This is not to belittle the important social relationships between local residents and the Service community. These will be studied throughout the narrative.

The history of Service aviation at Cranwell belongs to the twentieth century. There is a file in the Public Record Office[3] and the first entry on the minute sheet bears the tantalizing heading 'Ancaster—Site for Central Flying School'. It is a Director of Works (Air) file but its contents do not support the minute sheet's promise. There were probably two reasons why aviation came to the Cranwell area. Under Captain Murray Sueter the Air Department of the Admiralty had begun to establish a series of air stations round the east and south coasts of the British Isles, supplementing the coastguard system to alert the defences of our shores against sea and aerial invasion. The requirements for an airfield in 1913-14 were relatively simple. They were stated by Lieutenant-Colonel F H Sykes, Officer Commanding the Military Wing of the Royal Flying Corps.[4] 'The points which should be considered are the following: a. the inherent suitability of the actual locality itself for flying and military training; b. the geographical position of the station in the British Isles.'

The flat farmland including heathland around Cranwell was as suitable as many other possible sites for establishing an airfield. But there was a second factor at work. The Naval Wing of the Royal Flying Corps formed in 1912 had maintained its own flying instruction camp at Eastchurch on the Isle of Sheppey. One of its most competent airmen, Captain Godfrey Paine, RN was the Commandant of the Central Flying School at Upavon, Wilts, where military, naval and Royal Marine officers were taught to fly. The tendency of the senior Service to go its own way was confirmed by the creation of the Royal Naval Air Service on 1 July 1914. The outbreak of war found two separate Services developing their own training programmes but co-operating in a Central Flying School which quickly became inadequate to deal with the requirements of the Military Wing, now known generally as the Royal Flying Corps, in France. The Air Department of the Admiralty sought to put its own house in order. The brilliant but unorthodox Murray Sueter was replaced by Rear-Admiral C L Vaughan Lee as Director of the Air Department in 1915. The Board soon gave approval to a proposal to open a separate central flying school for the Royal Naval Air Service. This would mean the removal of naval pilot training from the Central Flying School at Upavon. The beginning of Service aviation at Cranwell was linked with the desire of the Admiralty to open a central training establishment for its pilots in a location which could support the new air stations being set up on the east coast of England.

Laying the foundations

Officers from the Admiralty, who included in their number Commander F C Halahan,[5] took part in a survey in late 1914 or early 1915 of likely aerodrome sites. The Aerodrome Selection Committee toured the countryside and decided to recommend the requisition under the Defence of the Realm Regulations of 'a large stretch of very flat country on the top of the heath above Caythorpe'. Cranwell was to be the site. What sort of setting, what community was thus to be transformed by the advent of the RNAS can be discerned in the pages of White's *Lincolnshire*, published in 1892. Cranwell in 1891 had 188 inhabitants and comprised 2522 acres of land. Sir J C Thorold was the impropriator, lord of the manor and owner of all the land except one farm belonging to St John's College, Cambridge. The manor had been held by his family for about three centuries and they resided formerly in a mansion which was pulled down in 1816, when a farmhouse was built on its site in Cranwell village. In 1891 Alexander Brimley Banks occupied the Hall Farm and Thomas L Thurlby occupied The Lodge as a farmer and grazier. Another important landowner of adjacent fields was the Marquess of Bristol. The Bristol family had come into possession of estates by a marriage to the daughter of Edward Carre (d. 1684), the last baronet of the knightly family of Carre, the munificent benefactors of Sleaford. In 1915 the owner of The Lodge Farm was Mr William Usher Banks, a tenant of Sir John Thorold. It was he who was to become central to the scheme of requisition of land.

The Admiralty acted quickly and issued an Office Memorandum on 13 December 1915. This stated:

> Captain Godfrey M Paine CB, MVO has been appointed as Commodore of the First Class in charge of the Central Depot and Training Establishment for the Royal Naval Air Service which is being formed at Cranwell. Instructions have been given to the Commander-in-Chief at the Nore to select a hulk in the Medway to be renamed *Daedalus*[6] and commissioned under that name. Commodore Paine's Broad Pennant will be hoisted in her for one day and will be subsequently transferred to the shore. The staff of the Establishment will be borne eventually on the books of HMS *Daedalus*. Commodore Paine has been authorized to correspond during the erection of the Establishment direct with the various departments of the Admiralty and to communicate personally with their Lordships if necessary. His appointment takes effect as from the 10th December 1915.

As in similar situations the Officer Commanding began to choose his team. Naturally he picked several who had already proved their worth at Upavon. Among them were the three officers who appear in the photograph (Plate 1) whose original hangs today in The Lodge at Cranwell. They were Flight Sub-Lieutenant F H M Maynard (subsequently Air Vice-Marshal F H M Maynard, CB, AFC), Flight Commander J H Lidderdale and Flight Lieutenant J R Potter.

Correspondence between the War Office and the Admiralty settled the date of opening for operations of the RNAS Central Training Establishment, Cranwell as 1 April 1916. The RNAS would then send no more naval pilots for training at Upavon. All naval personnel at CFS would revert to the Royal Navy and one half of the instructional machines, engines and models would be transferred from Upavon to Cranwell. The issue of General Memorandum No 14, dated 24 February 1916, was the official authority for setting up the new establishment (see Annex I). Cranwell later became known as HMS *Daedalus* but this, while attractive, is not strictly correct. All officers and ratings of the RNAS Central Training Establishment at Cranwell were borne on the books of HMS *Daedalus*. So also were other officers and ratings of the RNAS on other stations. *Daedalus* was the nominal depot ship for naval personnel not only on the east coast but also at Plymouth (Cattewater), Mullion, Tresco (Scilly Isles) and Newlyn.[7]

The Admiralty took possession of Cranwell Lodge Farm on 23 November 1915.[8] Mr Stone, an Admiralty lands officer, who found lodging at the Saracen's Head in Lincoln, was given authority to demand the requisition of land from various tenant farmers. He was also looking for a house suitable for the Officer Commanding. Roxholme and Leasingham Halls were inspected.

An account of the first arrivals at Cranwell has to be pieced together from four sources of reminiscence. The accounts agree roughly in outline but not in detail.[9] One day early in December 1915 a naval rating named Whitlock (later CPO W H Whitlock) was sent from RNAS Wormwood Scrubs to Grantham with a Lancia tender and told to put himself up there and to await orders via the stationmaster. He was detailed one morning to meet a Mr Marriott off the London train and to take him around, apparently to fix up contracts for supplies to the new camp. For several days Whitlock stayed in Sleaford. He visited Cranwell Farm, where many surveyors were at work. His next duty was to take three RNAS officers (Maynard, Lidderdale and Potter) from Grantham to the Bristol Arms Hotel in Sleaford. These officers had taken formal possession of the Cranwell Farm, ie, what is now the Commandant's house, The Lodge. On 16 December a CPO and thirty-five men from the Isle of Grain and Eastchurch were lifted from Grantham railway station by Foden lorries to their new billets at Cranwell. The men slung their hammocks in the various outbuildings and in the cottage in the grounds of the Lodge Farm. The remains of the sheds and stables can still be seen and the modern MT yard was built where several outbuildings stood. Conditions in winter were primitive and remained so for the next two months. It was decided that the Lodge should become the Commodore's residence. Work began at once to make roads and hard standings, to dig drainage channels and make the foundations for huts. The first sod was cut on 28 December 1915 and the construction of huts 1 to 12 began. A small army of navvies, including many Irish labourers, appeared. At this time the road from Byard's Leap to Cranwell village was a cart track used mostly by the farmer at The Lodge. All other tracks were rutted and pot-holed. Ratings and contractors' men pulled down stone walls to make roads over which the traction engines and wagons could pass. The weather was severe and the Director of Air Services in London acknowledged on 8 March 1916 that the Establishment would not be able to accept pilots for graduation until 1 April 1916.

In late January or early February Lieutenant Maynard was sent to collect the first aeroplane, a BE2c, from Chingford. He tells the story as follows:

> I duly collected the BE2c, No 3999, and set off for Cranwell. It proved to be too long a flight for this elderly aeroplane and I had to refuel. This I did by landing in a rather small field outside Bourne. The usual small crowd collected and when I had explained my needs, some kind chap rushed off and collected three or four two gallon cans of petrol from a local garage. With refuelling completed I set out on the last lap after a rather sticky take-off due to a railway embankment. I had a few feet to spare so all was well and I duly arrived at Cranwell to make the first landing there. The BE2c was housed in a canvas hangar.[10]

The reputation of several men was burnished during these difficult days. A driving leadership came from the Commodore. One of his ploys was to spur on the workmen laying water pipes by offering a five pound note to the foreman to exceed the allotted, flagged mark for the day's work. A water tower to store water from a source at Sleaford was soon marking the horizon together with several iron huts and hangars. An early association with the Lincolnshire Regiment was formed when, after dining with Commodore Paine, the Colonel of the 4th Battalion is said to have marched his men to Cranwell for useful trench digging practice.

By mid-1916 the camp was taking shape. There was a hospital in West Camp where some of the buildings have remained until this day. In naval fashion the RNAS ran its own bakery and butchery, buying on the hoof. It had a series of boiler houses and the whole camp was soon humming like a ship.

An officer of whom local people retain still a clear memory was the

First Lieutenant—Jimmy the One. He was Flight Commander J H Lidderdale.[11] Originally a paymaster, he went to Upavon as Secretary to Godfrey Paine. He learned to fly an aeroplane, took his RAC ticket on 13 January 1913 and became a flying instructor at CFS in November 1914. Paine picked him for Cranwell. His vigorous personality, patience and amazing powers of observation and memory helped him to produce order from mud and chaos and hutments and hangars from pasture and cornfields. His widow, Jill Lidderdale, described at the age of 91 in 1979 how she met her husband in Lincolnshire. Jill Walker travelled from London in a spirit of service and some daring to work on the land near Sleaford, baling hay for the horses of the Household Cavalry in France. The work was strenuous but she was a favourite with the traction engine drivers. It was while working in the fields that she encountered Lidderdale. The romance blossomed and the RNAS magazine *Piloteer* was proud to announce in the Christmas number 1916 that 'the marriage of Flight Commander J H Lidderdale to the youngest daughter of Mr and Mrs J Hansen Walker of Queen Elms Square Chelsea took place on Saturday 16 December at St Peter's Cranley Gardens London.' This was perhaps the first of many romances and marriages between members of the staff at Cranwell and local ladies from 1915 to the present day.

The acquisition of land

While the pioneer work was being done in Lincolnshire, the Chief Valuer of the Admiralty began moves to assess the claims of the various dispossessed tenant farmers. They had claims not only for land taken but for closure of roads, deprival of right of way, shooting rights and the use of quarried stone. The landowners concerned in the bargaining were Sir John Thorold, Lord Bristol's agent, and St John's College, Cambridge. The Treasury sanctioned the purchase of Cranwell Lodge Farm at 'the best price obtainable within limit'. Sir John Thorold offered to sell:

Cranwell Lodge Farm	990 acres
Woodlands	18 acres
Byard's Leap Gorse	$11\frac{1}{2}$ acres

but the Chief Valuer's reply to his asking price was much less than he expected. Mr Usher Banks too was offered less than half of his asking price for loss of profit and tenant's rights. Six cases were heard on 16 November 1916 under the procedure of the Losses Commission to deal with tenants' claims on loss of property and interests. The claimants were W U Banks and J & G Sardeson, tenants of Sir John Thorold, and Messrs G J Pilkington, F Ward, J Moore and E S Tomlinson, tenants of Lord Bristol. A total of 2201 acres is mentioned in contemporary records[12] as the land acquired but a figure of '2500 acres (roughly)' was given by the Superintendent Valuer of the Eastern Division of the Civil Engineer-in-Chief (Air) in a letter dated 16 August 1916.

Additional land requirements developed from the laying of the Sleaford to Cranwell railway. This was begun after the experiences of the hard 1915-16 winter to make a more reliable means of transport for bulk supplies and heavy equipment for the Cranwell camp. At first it was simply a temporary track laid directly on the ground and undulating with it. The laying of a permanent track was sub-let by the Great Northern Railway to Logan and Hemingway Ltd of Doncaster. The survey was completed in January 1916 and the line was finally handed over to Air Ministry Works and Buildings in February 1919, although separate sections were used by the contractors as soon as they were built.[13]

The argument over a fair price for the land and loss of interests went on throughout the war years. Eventually the title to the land changed hands and the property was purchased by Deed dated 4 September 1919 (Lord Bristol and others to the Air Ministry) and by Deed dated 5 June 1920 (Thorold to the Air Ministry).[14] The interest of St. John's College, Cambridge, appears to have centred on 'a small triangular piece of land with one or two trees growing on it' but the Bristol estates were a major owner of land. Their name continues to be preserved at Cranwell today in the Bristol Plantation or Wood and in the names of roads to officers' married quarters, ie, Plantation Road and Bristol House Road.

Repercussions of the denial of rights of way were felt for some time after the purchase of the land and as late as 30 June 1922 the Commandant, Air Commodore Longcroft, was involved in correspondence 'about a footpath from Brauncewell which runs right across the centre of the Northern Aerodrome'. Eventually an alternative footpath was agreed for the use of the general public between Brauncewell and Byard's Leap.[15]

RNAS Operations and Training 1916-1918

Cranwell was formed as a flying training station and through the decades since 1916 it has retained that essential character. As the

Admiralty General Memorandum No 14 of 24 February 1916 stated: 'All officers and men under instruction in aeroplanes, kite balloons and airships will be sent to Cranwell for a finishing course and for graduation.' Pilots would train up to the standard of flying 'a moderately fast tractor' at four preliminary training schools—Eastchurch, Eastbourne, Chingford and Redcar—and advanced training would be carried out at Cranwell. After graduating at Cranwell as aeroplane pilots, officers as a general rule were appointed to Freiston (some five miles east of Boston on The Wash) for about fourteen days' practical work in gunnery and bombdropping. Officers selected to qualify as seaplane pilots were appointed after Cranwell to seaplane stations such as Killingholme. Commodore Godfrey Paine in a letter dated 12 May 1916 to the Director of Air Service at the Admiralty proposed that Cranwell should train five hundred pilots in a twelve month, ie, approximately ten pilots per week. Accommodation, aircraft and all supporting facilities were therefore to be geared to a pupil population of one hundred pilots. The advanced training consisted of cross-country flying, flying fast scouts and fighting machines, firing from the air, bombdropping, wireless telegraphy and aerial photography. The initial requirement for machines was given as: 5 Maurice Farmans or Box Kites, 25 Avro or Curtiss, 20 BE2c, 10 × 1½ Strutters, 10 Bristol Scouts, 10 Nieuports and 10 others such as RE8 and FE2B.[16] About half were dual control aircraft.

Information from official sources about the early days at Cranwell being scanty, one must turn to the valuable *Piloteer*, the tuppenny magazine issued fortnightly by a sub-editorial committee of petty officers and an air mechanic under the direction of the Editor and Censor, the Rev J R De Vitre, RN. In issue No 4 the Sports Committee consisted of representatives of some twenty sections on the station—including the Boys. By late 1916 Cranwell was already a full equipped and established air station and included an Electrical and Wireless School in the eastern half of the station. Another permanent characteristic of Cranwell was implicit from the the outset, ie, an eastern and a western part of the station with the East Camp though not called that until much later being given over to a ground training unit.

Life on the Station

As training expanded so did the social and domestic activities at Cranwell and in the neighbouring communities. The church at Cranwell has always been a centre of interest and on 20 August 1916 a church for the station was dedicated by the Bishop of Grantham in the Gymnasium building. The village of Cranwell was from the outset closely associated with the camp. Its churchyard became the resting place for the many casualties of flying accidents. The first accident saw the death of Squadron Commander I H W S Dalrymple Clark who was formerly at CFS with the Commodore. He was killed when his machine crashed in cloud into that flown by a fellow instructor. There were to be 48 other casualties at Cranwell before Armistice Day 1918. After Dalrymple Clark the next two to die were a Canadian and a New Zealander.

The *Piloteer* chronicles the busy life of the station and the way in which the local gentry and townsfolk soon became involved with the sporting, theatrical and other domestic events at the camp. Sleaford was very much in bounds to personnel of the station and petty officers and ratings went ashore on the liberty boat, ie, a Foden lorry or later the 'Cranwell Express'. When a liberty train returned to Sleaford at the end of the Christmas and summer block leaves the organization was a little complicated. The men were divided into three parties—those who went into town, those who caught the first train back and those who marched some two miles out of Sleaford. By the time they reached that point the train was back from Cranwell to pick them up.

Mr J W Robinson established his Post Office in October 1916 on the south side of the road from the modern College. His Post Office was close to the huts which in RNAS days were Mess Decks where personnel of the station took meals. These huts later became the Cadets' Mess until the modern College was built. The naval hospital built in 1916 has retained its site and some of the original buildings until the present day.

A Cranwell Orchestra, Cranwell Brass Band and a highly talented concert party called the Beaux Bells helped to increase the good opinions which local folk were forming of the naval airmen at Cranwell. Names like Culverthorpe Hall as the venue for a Garden Fête and Aquatic Sports begin to appear in the *Piloteer* starting an association with friendly neighbours like General R L Adlercron and his family. The officers tended to go further afield for their 'nights ashore'—the delights of Nottingham and Lincoln were favourites. Mr A H Mather JP of Boston, whose parents kept the White Hart Hotel there, recalled many evenings in their private parlour where officers from Cranwell and Freiston would relax round the piano. The Beaux Bells were much in demand locally, including as they did such performers as Alec McGill, the pianist, and Harry Champion and at least once, as a guest star, Dorothy Ward, whose brother was stationed at Cranwell.

The flights over Skegness of the Freiston detachment became a daily exciting feature. One noted aviator from Cranwell used to perform aerobatics over Boston and finish his performance by dropping a box of chocolates on the end of a miniature parachute to be taken to a certain schoolmistress in the town.[17]

Unusual additions to the strength at Cranwell were provided by a detachment of Wrens who were quartered in the huts of West Camp, which became the Cadet Squadron huts after 1920. The second attachment was of some American Navy personnel, who were based at Cranwell between November 1917 and March 1918.

Commodore Godfrey Paine, who had got the air station under way, was appointed First Air Lord and Fifth Sea Lord at the Admiralty early in 1917. He was succeeded by Commodore J Luce, who was a sailor rather than an aviator in his predecessor's mould. Luce remained at Cranwell until April 1918, when he was succeeded by Brigadier-General H D Briggs.

By the time of the Armistice in 1918, Cranwell had become a key station in the great programme of flying training stimulated by the demands of war at sea and on the Western Front. It had grown so rapidly that HM King George V could write in his diary 'We motored to Cranwell which has now become the largest aerodrome in the world'.[18]

Lighter-than-air Activities

The activities of the lighter-than-air section, which was disestablished after World War I and whose history is not well recorded, deserve a more detailed description. General Memorandum No 14 of 24 February 1916 stipulated that: 'On and after 1st April 1916 all officers under instruction in aeroplanes, kite balloons and airships are to graduate as pilots at Cranwell.' Considerable accommodation and works services were required on the site across the north airfield. The section was placed to the east of the track, now a metalled road, which goes across that airfield and branches north-eastward. The name of the road leading to the modern married quarters also commemorates the section. Occupants of the married quarters well recall the old airship section when deep digging in their gardens strikes concrete. A portable kite balloon shed was erected in March 1916 and the first ascent at Cranwell was made in a kite balloon on 4 April 1916. Free balloon training was also undertaken in six coal-gas balloons which were brought up from Wormwood Scrubs. A portable SS shed was blown down in the course of erection but was completed early in November 1916.

A coastal airship shed was erected to house the airship C13 and early in 1917 the great double rigid airship shed was finished. Plans and approval to build a second rigid shed were cancelled in May 1917. Offices, a fabric hut, a compressor house, hydrogen and water-gas relief holders and a silicol plant completed the inventory of buildings of a tidy section. There were five officers and 131 ratings on the books in the first year, 93 of the ratings being under instruction. The first Officer Commanding met an unfortunate end which is described below. His successor for the duration of the war was Wing Commander J N Fletcher, whose qualifications for the post were impeccable. Commissioned into the Royal Engineers, he served at the School of Military Engineering, Chatham, and then with the Air Battalion, RE, at South Farnborough. He was one of the first to transfer to the RFC having taken his Balloon, Airship and Aeroplane Certificates. In 1914 he was appointed Captain of No 4 Airship and transferred to the RNAS as a Squadron Commander. He came to Cranwell after other airship commands and a spell in the Air Department of the Admiralty. He was a very keen officer and one of the old Air Department files from the Hotel Cecil[19] contains some vehement criticism by Fletcher of the slipshod training system at Wormwood Scrubs. The intervention of the Commandant at Cranwell resulted in the transfer of initial balloon training of airship pilots from the Scrubs to Roehampton. Cranwell was already setting high standards in training.

A list of the balloons and airships which operated from Cranwell in World War I is given at Annex III. Particular reference must be made to one sad incident which illustrates the hazardous nature of lighter-than-air operations. On 20 July 1917, SS39 was being walked back to its shed after an instructional flight, the wind being no more than 4 to 5 mph. An upcurrent of extreme violence caught and dragged her out of the hands of the landing party, who nearly all let go. The Officer Commanding, Wing Commander C M Waterlow, PO M G Collins and Aircraftman II S Lightstone held on to their ropes and were carried upwards to a great height. All three eventually dropped off the ship and fell to their death. SS39 made a free balloon landing about one and a half miles east of Cranwell village and was ripped but otherwise suffered little damage.

In the second year of training the complement of the lighter-than-air-section rose to 47 officers and 203 ratings. Seventy-five officers graduated as airship pilots in the year 1917-18. But despite this important contribution to the war effort and in spite of the special romance attached to lighter-than-air operations, the future at Cranwell as elsewhere was for heavier-than-air craft.

2 *Connections with the Royal Family*

The station formed at Cranwell by the RNAS and continued by the Royal Air Force after 1918 has always carefully treasured its associations with the Royal Family. As early as summer 1916 King George V and Queen Mary paid a visit to the newly established training centre of the RNAS. While visiting Army training centres and weapons firing ranges in Lincolnshire, the King and Queen motored across from Belton, where they lunched with the Brownlows, to Cranwell. 'Commodore Godfrey Paine showed us round. We saw any number of machines in the air and all the latest things in wireless, wireless telephones and for dropping bombs. A lovely day, quite hot. Left Leadenham at 4.30 and reached B.P. at 7.0.'[1]

The next Royal visit had an even more personal significance for in 1918 the King and Queen came to see their second son, Prince Albert, who was an officer at the station. Prince Albert, like his father before him, was originally intended to be a career naval officer. He was educated at Osborne and Dartmouth, but from his early days experienced frequent bouts of gastric disorder. The cause was finally and correctly diagnosed in 1917 and an operation was carried out on 29 November 1917. Prince Albert had already seen action at the Battle of Jutland in HMS *Collingwood* and prior to his operation had been serving in the battleship HMS *Malaya*. It was decided that his physical condition made it advisable for him not to return to the rigorous life of a naval officer at sea. He was therefore 'transferred from the Navy that Floats to the Navy that Flies and his appointment to the Royal Naval Air Service Station at Cranwell was gazetted on November 13'.[2]

The Prince received his appointment to HMS *Daedalus* on 1 January

1918 and reported there a month later. He was accompanied to Cranwell by Surgeon Lieutenant-Commander Louis Greig. The latter had known the Prince at Osborne and had served with him at sea, lately in the *Malaya.* A former Scottish Rugby international cap, Greig was a great athlete and had become a close friend and companion to Prince Albert. Greig was given the appointment of OC The RAF School of Physical Training at Cranwell. The Prince was appointed initially as Officer in Charge of Boys and later as OC No 4 Squadron, West Camp. He wrote to tell his father that there were already 500 RNAS boys in training and the numbers were going up to 1500. His reactions were very frank. He wrote to Queen Mary: 'My rooms are very comfortable in this small house, and I expect to use them a certain amount, as there is little room in the mess ... Most of the officers are very nice though a curious mixture of people in every walk of life.'[3]

It is interesting to note that his writing paper was headed 'RN Air Station, Sleaford, Lincolnshire' and not HMS *Daedalus.* The little house referred to in the Prince's letter of 6 February was the building near the present York House Officers' Mess at Cranwell and called York House.

Wing Commander A E F McCreary then a junior officer recently promoted from the ranks was second in command to Prince Albert in No 4 Squadron. He recalls in his memoirs, not published, being allowed to use the royal bunk in York House when he was Officer of the Watch. Prince Albert, following the King's wishes, was treated very much as all other officers on the station were treated and took his turn on the duty roster. McCreary confirms that Prince Albert was a keen and energetic officer. He was a stickler for discipline and routine. He wrote to his father on 9 February 1918, 'The life here is very different to a ship. Everything I deal with is a long distance from one another. As I think I told you I am looking after the boys, of which there are 500. This week I have been taking over from Mackay, a flight sub who was in charge of them. I am going to run them as an entirely separate unit to the remainder of the men and I am known as the Officer Commanding Boys. I shall have to punish them myself and grant their requests for leave etc. At present I have not got a proper office but hope to get one shortly. The work is entirely new to me and I found it rather difficult to begin with but I shall get used to it. They live in small huts, 20 boys in each and these give me the most trouble as they won't keep them clean without my constantly telling them off to clean them out of working hours. The petty officers are not altogether very satisfactory. One finds a tremendous difference between them and the proper Naval

P.O. But with a little persuasion I hope to make them understand what I want.'[4]

The Prince found conditions quite different from the ordered and well disciplined world of a battleship. But his well known determination to carry through a task to which he had addressed himself is evident in his letters home. However, more trials and stresses were to come his way because on 1 April 1918 the RNAS to which he belonged was merged with the RFC to become a new third Service, the Royal Air Force. One of the consequences was the concentration of boys' training at Cranwell. The new brick buildings in East Camp were not yet completed and there was a prospect of going under canvas for the 300 or more RFC boys who arrived at Cranwell in mid March. In time numbers were planned to rise to 2500 boys in total. There was a degree of confusion at Cranwell as attempts were made to reconcile the different drill, uniform and practices, including language, of the two former flying services. The Prince wrote to his mother on 24 April 1918, 'Nothing very much going on here now as everything is in rather a muddle owing to the new routines etc. But still they will straighten out in time.' With the RFC boys from Halton came, of course, their officers, including one Major Routley, who was assigned to teach Prince Albert 'Military Law, drill etc which of course I know nothing about.'[5]

It is interesting to note that the Prince's writing paper at this time bore the manuscript amendment at the top which read: 'R Air Force Station, Sleaford, Lincolnshire.' Brigadier-General H D Briggs had meanwhile taken over in command from the last RNAS officer in charge, Commodore J Luce. He was appointed Air Officer Commanding No 12 Group, with Cranwell as headquarters.

The King and Queen were particularly concerned about the health of their second son and were grateful to learn that the strong air of Lincolnshire and the hard work were sending him to bed a tired but healthy man. He was riding a horse, playing some tennis, especially with Louis Greig, and learning to drive a car. He was taught on a chassis under test by 'Whiteley a Lieutenant Commander RNVR who is in charge of all the Transport.'[6]

Prince Albert (PA, as he became known to his seniors) impressed his fellow officers with his frank, modest and serious attitude towards his duties and life on the station. Group Captain Jimmy Jeffs, who was then Officer Commanding No 19 (RNAS) Squadron at Westgate, returned to Cranwell for a radio and hydrophone course in 1918. He told the author that the Prince was popular and well esteemed by all who came into contact with him.

The Prince made his first aeroplane flight. 'I went up in an aeroplane for the first time yesterday afternoon. A man called Pierce[7] took me up. It was a curious sensation and one which takes a lot of getting used to. I did enjoy it on the whole, but I don't think I should like flying as a pastime. I would much sooner be on the ground!! It feels safer!!'[8] The letter was sent to his mother, who probably appreciated the sentiments well, especially as other letters had indicated the not too rare occurrence of aircraft crashes in flying training.

The King and Queen took an early opportunity to see their son during a planned visit to Lincoln. King George V wrote in his diary:

> Thursday April 11th. In the train Skellingthorpe, Lincoln.
>
> We left at 9.50 and reached Leadenham at 10.20 where Brig Gen Briggs and Bertie received us. We motored to Cranwell which has now become the largest aerodrome in the world. Inspected the boys who Bertie is in charge of, visited the sheds and saw all the different kinds of machines, also the classes at instruction and we saw the airships in their hangars. Unfortunately it was misty and foggy and so we could not see any flying. We joined our train at Leadenham at 1.30 and left for Windsor.[9]

This was a great occasion for Prince Albert to show off his former RNAS boys while the late RFC boys were on parade under Major Routley. As the *Piloteer* noted in its report of the visit, the King was delighted with the reception given him and as he drove past the cheering troops remarked 'Good boys, good boys-that's the stuff'!

The Prince was very upset about the resignation of General Trenchard from the post of Chief of the Air Staff. 'Everything here as you may imagine', he wrote on 21 April 1918, 'is in a very unsettled state owing to the changes in the Air Board and nothing is settled yet as to what routine we are working under. We are now having a mixture of naval and military routine which is not a great success.'[10] The King replied immediately, 'I am sorry that General Trenchard's resignation has produced an unsettled state of affairs in the Air Force which is only to be expected but I hope nobody will lose their heads but will try and work loyally under the new Air Board. I know General Sykes is not popular in the Force but that cannot be helped; everyone ought to try and do their best at such a serious moment as the present. I hear General Godfrey Paine is going to remain.'[11]

In truth, 'officers and instructors were arriving and departing with bewildering rapidity and the already established system of training boys, which had been Prince Albert's special care and interest was completely disrupted. Having been in charge of all the boys he was not unnaturally annoyed at finding his command cut to a mere fraction, with newly appointed and inexperienced officers placed over him.'[12]

He grew somewhat depressed but from 1 May 1918 he was considerably heartened by his father's consent to his living with the Greigs in their cottage at South Rauceby. This enabled him to escape from the rather bleak camp at the end of a day's work and to enjoy the mundane tasks of gardening and looking after chickens in the happy home atmosphere in the South Rauceby cottage. The villagers still recall the sight of their future king living among them as a neighbour and cycling to work. He was indeed now in fine health and a fit young man. Cranwell had worked wonders for him. The AOC also ensured that he was kept in touch with important visitors and he took meals at The Lodge with Axel of Denmark on 13 February and with Sir William Weir and General Godfrey Paine on 11 June 1918. He was also carrying out other royal duties not normally associated with an officer on a training station. In his letter to the Queen on 12 June 1918 he noted: 'There is a fete at Sleaford for the Prisoners of War to which I am going and on Saturday I am going to Lincoln for a matinee for the journalists who have fallen in the War. Wigram advised me to accept it when I was asked. It is always a good thing to do things like that around here.'[13] The Prince was beginning to know his Lincolnshire folk.

Despite the arrival of Lieutenant-Colonel A C Barnby RAF to take command of the Boys' Wing and despite the gradual evolution of a new pattern of drill and technical training, the Prince was not really settled and happy. It was decided that he should be transferred to gain experience of RAF officer cadet training at St Leonard's-on-Sea and with Louis Greig he moved there on 1 August 1918.

Cranwell's association with the Royal Family therefore began from its earliest days and it has been maintained through the reigns of three monarchs. HRH The Duke of York, later King George VI, was the only member of the Royal Family until recent times to achieve the distinction of gaining his official pilot's certificate. This he did on 31 July 1919 at Croydon after tuition from Lieutenant W A Coryton, later Air Chief Marshal Sir (William) Alec Coryton. But his first flight was at Cranwell during his time there as a serving officer of the RAF.

3 *Early Years at Cranwell*

In the last year of the Great War, RAF Cranwell continued to turn out pilots urgently required for the final offensives on the Western Front. The station was now the home of Headquarters No 12 Group. At the Armistice on 11 November 1918 the establishment of the station was:

HQ No 12 (Training) Group

HQ 59th Wing

No 56 Training Depot Station

No 57 Training Depot Station

No 58 Training Depot Station

Airship Training Wing

Boys Training Wing

Physical Training School

The last three were under direct control of the Air Ministry for technical administration.

In May 1919 the Officer Commanding No 59 Wing was Lieutenant-Colonel Peter Portal later to become MRAF Sir Charles Portal, Chief of the Air Staff 1940-45. He was transferred in June 1919 to Headquarters No 12 Group, the nucleus of the future Royal Air Force College. He also took the opportunity, as did so many later serving officers, to marry a local girl. On 22 July 1919, Peter Portal married Joan Welby, the youngest daughter of Sir Charles and Lady Welby of Denton Manor.

The stormy exchanges between senior officials and officers in

London which had marked the birth of the new Service and had been the subject of Royal correspondence did not materially affect the conduct of flying operations in the closing months of the war. Trenchard after his resignation from the post of Chief of the Air Staff was eventually persuaded to take command of the Independent Force and later the even more grandiosely conceived Independent Allied Air Force in eastern France. Major-General Sir Frederick Sykes had been appointed by Lord Rothermere to succeed Trenchard and he was CAS when the Armistice came. The success of Lloyd George in the General Election of December 1918 and the reconstruction of his Coalition Government soon indicated that drastic reductions, especially in military expenditure, were to be made to get the country back to a peace-time economy. The Prime Minister invited Mr Winston Churchill to head one of the Service departments. 'Make up your mind whether you would like to go to the War Office or the Admiralty, and let me know by tomorrow. You can take the Air with you in either case; I am not going to keep it as a separate department.'[1]

In the event Churchill against his inclination was persuaded to go to the War Office and in addition was persuaded to heed the arguments of those who feared to go back to the chaos of 1917 before the formation of the Air Ministry. Reluctantly the Lloyd George administration was converted to keeping the Air Ministry and the Royal Air Force in independent existence. But the price to be paid was a rigorous cutting of aircraft, equipment and manpower. To carry out such a spartan and ruthless exercise Sir William Weir told Churchill that Trenchard was the man he needed as CAS. 'He can make do with little and won't have to be carried.'[2] In March 1919 Trenchard was re-appointed as Chief of the Air Staff and Sykes became the first Director-General of Civil Aviation.

Throughout 1919 Trenchard, after weeks of serious illness, was concerned with drafting and negotiating what were virtual blue prints for the shape of the emergent Royal Air Force.

The most significant of his papers was entitled 'Permanent Organization of the Royal Air Force'[3] which was accepted by the Secretary of State for War and Air in December 1919. Paragraph 6 of that paper dealt with the cadet college which was to be the main channel of entry for permanently commissioned officers. A Reconstruction Committee under Brigadier-General T I Webb-Bowen CMG had been considering the age of entry for candidates for commissions in the Royal Air Force and had made recommendations[4] in line with the entry conditions for

Royal Navy and Army academies. But Trenchard insisted in a memorandum written in September 1919 that 'the Royal Air Force should be maintained as separate from the Navy and the Army.' He disposed of the arguments that future RAF officers should be trained in existing RN and Army cadet establishments such as Dartmouth, Woolwich or Sandhurst. His most telling and typically pragmatic argument was that the training of RAF cadets at the RMC or any other existing naval or military college would prevent their being taught to fly during their training owing to the absence of aerodromes and unsuitable surrounding country.[5]

Trenchard instructed his training staff, in particular Colonel E R Ludlow-Hewitt, to draft notes for the Secretary of State outlining training plans for officers, boy mechanics and aircraftmen of the Royal Air Force.[6] The sections on the 'Training of the Permanent Officer' are interesting. The sources of supply are said to be:

a. Entrance from public and other schools through a competitive examination held under the auspices of the Civil Service Commissioners, and by special nomination. The regulations governing the entrance of these Cadets have been co-ordinated with those governing the entrance of cadets to the Royal Military Academy, Royal Military College and Naval Training establishments.

b. Entrance from Universities.

c. Promotion from the ranks.

d. Entrance through the Boy Mechanics Training establishments.

On leaving the college after a two-year course it was envisaged that the cadet 'will have a general although somewhat elementary knowledge of the Navy, Army and Air Force—will be a first class pilot of the Avro—will have an elementary knowledge of the work carried out from flying machines and will have a solid grounding in the duties of the mechanics of an aeroplane squadron.'

Ludlow-Hewitt proposed that post-Cranwell cadets should spend four months with the RAF which was working with the Navy and four months with the RAF working with the Army. But the Director of Training, Philip Game, thought little of this. 'The esprit-de-corps of the RAF would suffer. They [the ex-cadets] should do four months as a Gunner or Wireless Operator in one of the chosen branches of the RAF and concentrate on improving their flying skills.' This strongly practical note was characteristic of most early thinking about the role of the RAF College and took its tone from the Chief of the Air Staff himself.

The content of the training syllabus, especially the education of cadets, was the subject of many discussions and advice was sought from reputable quarters. While Ludlow-Hewitt and Colonel Curtis of the Educational Service canvassed opinions from existing training schools such as Halton Park and Flower Down, a Committee was formed under the chairmanship of Lord Hugh Cecil to consider the Preliminary Education of Candidates for Royal Air Force Commissions. Its terms of reference were as follows: 'To consider and report how far a special system of Education is necessary for candidates for RAF Commissions and in the event of such a special system being necessary, what form it should take, at what age it should begin, and whether the Government should give financial assistance, and if so, in what manner.'[7]

The Report produced contains some penetrating and imaginative observations on the qualities required of an officer in the Royal Air Force. Its general conclusions might be considered still valid. 'First, that the doors of the Air Force should be thrown open to boys of sufficient physical, mental and moral capacity; secondly, that they should be trained for the Air Force rather by general culture than by instruction in special technicalities.' There can be few reports which draw on passages, to delineate the elusive qualities of a gentleman, from Burke's Reflections on the Revolution in France, Cardinal Newman's Idea of a University and Lord John Cavendish's Speech proposing Mr Thos Townshend for Speaker in 1770.

In short, Trenchard sought to ensure that the RAF Cadet College should be founded on the best principles of education and instruction. He organized a conference in the Hotel Cecil on 2 July 1919 and entertained his visitors to luncheon at the Savoy Hotel nearby to put them in a receptive mood. After luncheon they were briefed on the draft regulations proposed for the new college. Among the distinguished guests were the Headmasters of Wellington and Haileybury and the First Commissioner of the Civil Service. The case for the establishment of a Cadet College at Cranwell in Lincolnshire was sent to the Treasury on 15 September 1919. Mr Stanley Baldwin rang the CAS on the telephone on 2 October 1919 to tell him that the Treasury had given the go-ahead for the scheme.[8]

Cranwell Air Station came under command of Air Commodore C A H Longcroft on 1 November 1919 and he was appointed Commandant of the RAF Cadet College and AOC of Cranwell Air Station on

3 November 1919. Wing Commander A L Godman CMG, DSO was nominated as the administrative officer on his staff. Instructors and men who had arrived back from overseas were to be specially selected at Manston for service at Cranwell.[9] The next few months saw interchanges of letters between Cranwell and the Air Ministry as vital problems of staffing were met. In particular the difficulty of obtaining the right type of men as butlers, stewards, waiters and servants was very great. So too were the arguments with the financial branches about the purchase and maintenance of suitable clothing for their duties.

An initial establishment for four squadrons of up to 50 cadets per squadron was eventually drawn up. The figures are interesting. It was proposed that eventually the strength of the unit should be:

56 officers and 516 airmen

200 cadets

1 Headmaster

1 Schoolmaster, Grade 1

7 Schoolmasters, Grade 2

There were to be 1 Head Steward (Mr J Simpson) 4 Under Stewards (including Mr E Curt), 4 Carvers, 8 Silvermen and 35 Servants.

The MT strength was to be 1 Laundaulette, 1 motor car, 8 heavy tenders and 8 light tenders. As early as 24 November 1919, the Commandant was writing to Philip Game at the Air Ministry 'I am trying to get the Blankney to meet at the Mess on Saturday, December 13th and will fix you up with a hireling.' Cranwell was taking shape and Air Commodore Longcroft was no doubt bucked to receive a letter from Trenchard dated 4 November 1919 which read: 'I am very glad indeed to hear that Cranwell is going to be such a great spot. I hope you will find me a horse and give me a day's hunting some time about Christmas, when you really get going.'[10]

The RAF College at Cranwell was opened on 5 February 1920. The first course was attended initially by 52 cadets of whom two were sub-lieutenants and 15 were midshipmen transferred from Dartmouth. The balance was made up by the first School Entry.[11] The College Adjutant was Flight Lieutenant C G Burge OBE and the civilian staff were S P B Mais, Professor of English, F G Swan, Instructor in English, B A Smith and G J Pytches, Instructors in Mathematics and Science. All four were Cambridge or Oxford graduates. Two squadrons, A and

B, were created and the flying training wing was commanded first by Squadron Leader C F A Portal, DSO, MC who was soon afterwards succeeded by Wing Commander L W B Rees, VC OBE MC AFC. A message from the Chief of the Air Staff spelled out clearly the mission of the College. 'We have to learn by experience how to organize and administer a great Service, both in peace and war, and you, who are at present at the College in its first year, will, in future, be at the helm. Therefore, you will have to work your hardest, both as Cadets at the College and subsequently as officers in order to be capable of guiding this great Service through its early days and maintaining its traditions and efficiency in the years to come.'[12]

The selection process for entry to the College was outlined in FS Publication 121. Entry from schools was by way of a written examination arranged by the Civil Service Commission. The subjects were in two classes. Class 1 (Obligatory) consisted of four papers each marked out of 2000. The subjects were: English; English History and Geography; Mathematics (Elementary) and one of seven modern languages. In addition not more than three subjects could be taken from Class 2. These subjects comprised: Latin, Greek, one of the seven modern languages not taken as an obligatory subject, Mathematics (intermediate), Mathematics (higher), Science (Physics and Chemistry), Elementary Engineering. Candidates were required to obtain not less than 30 per cent of the total marks allotted in each of the obligatory subjects. Since the same examination was taken by candidates for the Royal Military College, Royal Military Academy and Royal Navy (Special Entry) cadetships, candidates could opt for first and second choices of cadetship depending on their examination results.

The age limits for entrants were set between 17½ and 19 years and candidates were expected to pass not only the written examination but also to face an exacting Interview Board and a full aircrew medical examination in London. Trenchard at one time thought that selection could be merely by interview and by nomination but Lord Hugh Cecil and others urged him to adopt more objective criteria. Privately, Trenchard was reluctant to lose any boys in whom he thought he could discern future potential.

The fee for the written examination in the early days was five pounds and parents were expected to pay a fee for normal cadetships not exceeding seventy-five pounds per year. They were also expected to pay thirty-five pounds before entry and thirty pounds at the beginning of a cadet's second year towards the cost of uniform and books.

These were considerable sums for those days. Trenchard knew that he was asking parents and schoolmasters to support a venture completely new and unproven and his attitude and that of his staff towards the College in its first years reflected a burning interest and concern that often must have inhibited and disturbed the Commandant and his staff.

The curriculum to be offered reflected the high-minded and liberal views of Lord Hugh Cecil's committee. The subjects to be studied in the first year were: English Language and Literature; The British Empire; Practical Mathematics including Mechanics and Draughtsmanship; General Elementary Science; History of the Royal Air Force; Outline of the Organization of the Army and Navy and characteristics of the various Arms and Types of Ships; Map Reading; Drill (with Rifles) and Physical Training; Hygiene and Sanitation; Practical Work in the Workshops (in wood and metal); The Morse Code; Passenger Flying incl. instruction in Map Reading, and use of Compass and Machine Gun. A somewhat indigestible mass of information which soon became subject to various syllabus revisions. The second year promised more to cadets who wanted above all to get into the air and to fly. The second year subjects were to be: Theoretical and Practical Instruction in Internal Combustion Engines with Magnetos and their Management; Theory of Flight; Practical Instruction in Rigging; More Advanced Work in the Wood and Metal Workshops; Outline of Wireless Telegraphy and Telephony; Machine and Lewis Gun; Instruction in Aviation. Cadets could drop a term if they failed to reach the required standard in examinations or could be liable to removal from the College course.

Trenchard was conscious above all else of the necessity of attracting candidates of good quality. He had asked the Treasury for approval to award 15 per cent of places on each entry as King's Cadets, ie, with fees, uniform and book expenses paid. But the Treasury would agree only to the award of five King's Cadetships.[13] However other scholarships were endowed, in particular the Wakefield and Victoria League Scholarships. Sir Charles Wakefield[14] provided funds for a period of three years for the annual award of two scholarships each to the value of seventy-five pounds. Preference was to be given to the sons of parents whose reduced circumstances were due to the late war. The Victoria League Scholarships were financed from a gift of £1500 given to the Air Council by the Orange Free State Branch of the Victoria League. Two scholarships were to be offered for competition, one for each entry, to candidates whose fathers were British subjects of European descent and normally resident in the Union of South Africa. This scholarship unlike the Wakefield Scholarship could be held in addition to a King's or a Prize Cadetship.

Pay for cadets was five shillings per diem increasing to ten shillings per diem for the second year of residence. The Treasury had agreed to a daily allowance of one shilling per cadet towards meeting the cost of washing, messing and so on. No free issue of rations was to be made to cadets and daily allowances were payable only during College terms.

Life at Cranwell therefore promised to be novel but somewhat spartan and not inexpensive for those who first entered the Camp in 1920. What sort of place were they going to? They were going in fact to a camp which had not changed very much from the RNAS Station of 1915-18. The two squadrons A and B were to be housed in three lines of wooden and iron huts to the south of the road through the Camp from Byard's Leap to Cranwell Village, ie, in West Camp. The huts were linked together by covered ways to give some protection from the rain, snow and chill winds from the North Sea. A galvanized iron building known still as Mess Decks was to be the Cadets' Mess. It had been the dining-hall for RNAS ratings. The Mess contained at the back a long dining-hall with a curtained alcove for the band on guest nights, an ante-room (later the tea room) and a Fancy Goods Store. In the centre was the lounge hall fronting the kitchens and at the front three ante-rooms (one turned into a billiards room) and the library. The entrance faced the main road; between the Mess and the road were two tin huts—the West Camp guardroom and Mr Robinson's post office. To the rear of the Mess was a large parade ground and on its eastern side the Drill Shed surmounted by a clock, the arbiter of cadets' punctuality on parade or defaulters. Five flight cadets were quartered in each hut which comprised a dormitory, toilet facilities and a sitting/study room. Only in the Senior Term did cadets attain the privilege of separate cubicles. The Officers' Mess, again a legacy of RNAS days, was an iron-built building on the site of the present York House Officers' Mess. There were very few married quarters. The Lodge and the Assistant Commandant's bungalow were almost isolated features. The aircraft hangars on the South airfield and the Station Sick Quarters in 1920 were on the sites they still occupy. Both the North and South airfields were grassed. On the north side of the road through the Camp, where the new College buildings and the Orange are now, there were various huts for instruction and near the road were three large water towers, various hangars including later the famous Hangar Church, and such sheds as the indoor rifle range, instructional huts and the dope shed.[15] East Camp was almost alien territory, the home of the Boys' Wing, where huge new brick built barrack blocks were just being completed in 1920. Across the North airfield along Lighter-than-Air Road the gaunt vast Airship Shed stood as a reminder of airship and balloon activities in the Great War. As one contemporary witness wrote: 'It was my first visit to Lincolnshire and my first impression was that Cranwell was more remote from civilization than Mullion . . . The camp itself was extremely bleak.'[16]

The Air Estimates for 1919-20 tell their own story. Under Works Services is a vote heading: 'Alteration to huts and mess deck for Cadets' accommodation—£18,000.' Trenchard indeed was making do with little. The school entry in civilian clothes and bowler hatted, having obtained the necessary quota of marks in the entrance examination and having passed the Interview and Medical Boards would arrive by special carriage from King's Cross at Sleaford station. From there either by the Cranwell Express or in lorries with solid rubber tyres with canvas overhangs they would be driven to Cranwell, greeted there by drill sergeants and allocated to squadrons and their huts. Uniform from one of the prescribed London tailors would not arrive until term was much older. The spectacle of cadets in civilian clothes presenting arms was one not easily forgotten by the older cadets.

The life of the cadet in the early days was tough and demanding. The day began about 0645 hours. Each morning at 0800 hours there was a Colour Hoisting Parade. On Saturday morning there was a Station Parade with the Commandant on horseback and on Sunday there was compulsory Church Parade. On Station Parades Cadet Wing would be formed up facing the Drill Shed with Flying Wing on its right and the Boys' Wing on its left. After daily parade there was usually a drill period and then cadets were marched away to lectures, workshops, more drill or other activities until 1230 hours. Between then and 1400 hours was a luncheon interval. But from 1400 hours to tea-time at 1630 there would be more organized activities. After tea there was study preparation. Cadets dined-in every night in mess kit and once a week normally on Wednesday there would be a Guest Night with officers present. One cadet of the early days wrote: 'There were more opportunities to play games than we had time to play, rather more work than we thought we could do, the whole leaving us barely sufficient time to tie our GS tie for dinner.' Cranwell started off in the way that most ex-Cranwellians would probably testify that it has carried on! There was little time to spare. Group Captain F E Nuttall[17] told the author, 'I came to Cranwell because I wanted to fly.' When he was asked about his social life there, he said, 'There were no girls in my life—but they did come to the end of course dance.'

For four terms, almost two years, this was a strongly male society. The cadets had their own disciplinary structure. Each squadron had an under officer, two sergeants and two or more corporals. It was they who saw to the correct code of behaviour and taught new entrants the way to conduct themselves. Some of the traditions and practices were ritualistic and by no means gentle. One form of trial for first termers was the singing of a song in a public room. Those who had the misfortune to incur disfavour for whatever reason had to endure the ordeal of 'creeping to Jesus'. They followed a trail of pepper while blindfolded and scantily clad until they were tipped outside through an open window and doused with a fire bucketful of cold water.[18] Among other more formal trials cadets had to learn to change clothing with lightning speed especially if they had to appear on defaulters with immaculately wrapped puttees.

The Squadron Commander and his Flight Lieutenant were of course responsible for major disciplinary decisions but the internal working of the College was very much in the cadets' own hands. Officers attended dining-in nights but in those days the gulf between officer and cadet was wide. Even in the Flying Wing relations were much more formal than is now common. The flying instructors were not treated as godlike creatures but they were the men who held the key to the uncharted element of the air which the cadets wanted to master. There were originally A, B and C Flights and the workhorse was the Monosoupape rotary engined Avro. Instructors had the use of Bristol Fighter, DH9A and Snipe aircraft usually from the North airfield but for the cadet it was the Avro painted in its distinctive flight colours.

For a variety of reasons the actual flying instruction which cadets could obtain was not entirely satisfactory in the early years. F E Nuttall said he amassed only 56 hours 5 minutes flying in two years and during his first year had only 8 hours 45 minutes of passenger flying. This lack of air time may have been the cause of some disgruntlement and apathy towards other aspects of the College syllabus. Cadets were encouraged in workshops and by their NCO instructors to become very practical about engines and airframes. As part of their training Phelan and Moore (P and M) motor cycles stripped down to parts were issued to all cadets, who had to assemble them. After achieving this feat, cadets could use them for their own recreation. They could draw spare parts from stores using proper accounting procedure and could draw a free tankful of petrol each week. Motor cars were not allowed but motor cycles were. Cadets were virtually confined to camp and symbolic spiked iron railings surrounded the West Camp site. Sleaford was out of bounds; it had been a local pleasure spot for the RNAS. Only the Senior Term were allowed to go into Lincoln.

But life on the station was strenuous and every day was a full one. Cadets appeared to find the existence not only tolerable but rewarding. Four terms brought cadets to graduation. But not to their wings or indeed a very high standard of airmanship. Aircraft then were equipped with a compass and a bubble indicator for straight and level flying. That was about all. A cross-country flight to another airfield was made on a time and distance plotted course on a fixed bearing. Navigation was necessarily rudimentary and no night flying was attempted. It was on a pilot's first squadron after the completion of certain checks and tests that the ex-Cranwell officer received his wings. There was no passing-out parade as such but usually a talk by the Commandant or a distinguished senior officer given in the old Gymnasium. F E Nuttall remembered clearly the admonition of his Commandant, Air Commodore C A H Longcroft, 'Above all take care of your men—this is your first priority.' The influence of RNAS and RFC training and traditions was still strong.

The eagle eye of Trenchard missed nothing—even from London. On 23 March 1920 HRH Prince Albert, accompanied by CAS and Wing Commander Louis Greig, returned to visit the station he had known in 1918. This must have kept the Commandant very much on his toes. Then again on 19 April the Under-Secretary of State for Air, Lord Londonderry, paid a formal visit to the College. The Commandant was not spared written advice. On 15 January 1920 before the College was officially opened Trenchard was writing to Longcroft[19] The Government will build a Church in due course; there is no doubt about that, but it will not be at present. I do not altogether agree with you that the services cannot be held in the gymnasium at present.' And again on 10 March 1920 Longcroft was asked 'Who have you got up there who can train the boys in Rugby Football? After all this is the best game for making an officer and a gentleman out of any material. If we want to do well in the Air Force, I believe that Rugby is the best game to help us. The best schools for Rugby are places like Dartmouth,

Sandhurst and Woolwich. Rugby football should be encouraged as much as possible next winter.' Press comment and feed back from schoolmasters was grist to Trenchard's mill and on 10 September 1920 he was again writing to the patient Commandant. 'I was annoyed to see a long description in the Press about Cranwell. All of you are spoken of by wrong titles and Cranwell is described as if it were an annexe of the War Office.'

But Cranwell had been launched and measures to improve it in many respects were being taken. For example, the Medical Board report on candidates for the first entry brought out the facts that many candidates were being persuaded to put Woolwich or Sandhurst before Cranwell as first choice because mothers had a fear of flying. Secondly candidates who were otherwise excellently equipped were being turned down for Cranwell because of the high standard of visual acuity which was demanded. An American applicant for entry produced the ruling that only British subjects should have the right of entry to Cranwell.[20]

The Secretary of the Air Ministry also deemed it advisable to write to the headmasters of most public schools in September 1920 to point out that the RAF was now an independent Service.[21] It had an independent Service academy and the pay and prospects for school entrants were good. He was at pains to point out too that the unsatisfactory flying training conditions of the RFC training camps in World War I had been replaced by much more gradual training, ie, it was safer and the number of accidents would be very small in future years.

We have some extended testimony from one witness who was very much on the inside of the College in its first formative year. Mr S P B Mais, who had been a member of Lord Hugh Cecil's committee, had left his teaching post at Tonbridge School to become the first Professor of English at Cranwell. His account of Cranwell life given in his book *All The Days of My Life* is a bitter-sweet one. He acknowledged the enthusiasm and drive of the Commandant, Longcroft, 'who appeared to be obsessed by three passions—hunting, flying and Cranwell. He took me over the station with much the same air of eagerness that a motor-enthusiast would display while showing you his car.'

He also considered the officers a charming set of men, nearly all of them regular Service men, either from the Army or the Navy. 'There was a useful element of university-bred men.'

He was not as enamoured of the way the College was run and organized. Basically Mais felt that the cadets should be treated like undergraduates whereas the College authorities were keeping them under somewhat severe discipline, including the maintenance of bounds and roll call. Mais had to admit disillusionment over the cadets' reaction to his attempt to instill in them a love of English literature. His debating society and play-reading society fell flat. He did by hard work succeed in launching the first volume of the RAF Cadet College Magazine published by Gale and Polden Ltd in September 1920 and attracted to its columns articles by Lord Hugh Cecil, Sir Sefton Brancker, Sir Philip Game, a poem by Maurice Baring on Daedalus and an article by C A P (Portal) on 'The Cranwell Beagles'.

Perhaps because one section of the first entry had already a year's experience of disciplinary training in the Royal Navy and perhaps because what the cadets wanted was to get into the air as quickly as possible and resented other instruction as superfluous or inessential, Mais could but conclude 'I had lived among eighteen year olds for twelve years, and I have never known more delightful types than some of these Cadets; on the other hand I have never come across a duller, more conventional or more listless type than some others.' Perhaps it was because the methods of the barrack square were carried into every department of our lives.' There was a shortage of books, even of paper, and all the teething troubles of a new enterprise had to be endured. Mais was an energetic and unconventional schoolmaster and he ran foul of authority in a manner which was perhaps avoidable and unfortunate. Some cadets' essays criticizing the Cranwell system were placed into the hands of the Commandant. The upshot is best told in Mais' own words. 'It was owing to my passion for attending Meets that I lost my job in the RAF. The AOC had sent for me and I never got the message as I was out hunting. By the time I did see him the affair—which could easily have been settled by him—had passed out of his hands into the hands of Lord Londonderry who was not cognisant of the circumstances.'[22] Mais was replaced by Rupert de la Bere in April 1921 and de la Bere continued in post at Cranwell until 1938. But in his first volume of the College Magazine, Mais made a major contribution to Cranwell. He succeeded in getting the following inspiring paragraphs from Mr Winston Churchill:

It pleases me very much to think that the RAF Cadet College is to have a magazine of its own and I hope it will achieve the most complete success. The Chief of the Air Staff has already sent you a message, the sentiments expressed in which I entirely endorse. Nothing that has ever happened in the world before has offered to man such an opportunity for individual personal prowess as the air fighting of the Great War. Fiction has never portrayed such extraordinary combats, such hairbreadth escapes, such absolute superiority to risk, such dazzling personal triumphs. The devotion and courage of the airman was no higher than that of his comrade on the land or on the sea; nor were his trials or sufferings greater. But the battle to the death in the high air called forth a combination, for the purposes of war, of spirit, eye and hand more complete and more harmonious than any previously believed to be within the range of human nature.

It is to rival and no doubt to excel these feats of your forerunners in the Service that you are now training yourselves, and I, for one, look forward with confidence to the day when you who are now at the College will make the name of the Royal Air Force feared and respected throughout the world.

Lord Trenchard speaking to his biographer Andrew Boyle shortly before he died said: 'I thought they'd dislike Cranwell to start with, but hoped that eventually they'd appreciate my intentions. Marooned in the wilderness, cut off from pastimes they couldn't organize for themselves, they would find life cheaper, healthier and more wholesome.' Cranwell was far from the 'vice and pleasure area' of London and it was a bleak camp with few traditions. But those traditions were glorious ones from the RNAS and the RFC. Cranwell had one supreme attraction to the cadets; it was a marvellous place for flying. Whatever else it became or shall become Cranwell remains the home of primary flying training, a venue for young fit men with abundant facilities for healthy recreation.

4 *RAF Cadet College Life* 1920-1929

'Bliss was it in that dawn to be alive
But to be young was very heaven'

Memory casts a golden afterglow on events of one's youth. But in a very real sense the early years of the Cadet College were filled with novelty and excitement such as comes to innovators and experimenters. Life was simpler; aircraft were less complicated; the aim was clear and unambiguous. The Commandant was reported in the College Magazine of April 1921 as stating: 'Although an officer who could not fly was useless to the Air Force, yet the mere ability to fly by no means qualified an individual to become an officer in the Royal Air Force.' Strong stuff but the priorities were clearly defined. First came flying, then came physical fitness and discipline, sport and drill, and then came academic subjects especially those supporting the skills of aviation. Oral testimony from surviving members of the early courses at Cranwell is unanimous in its acclaim of the full life they led—mainly extrovert, fit young men proud of the new way of life they were exemplifying and eager to join the squadrons in the young Royal Air Force.

That invaluable publication[1] 'List of Graduates' begins to record many who subsequently have become household names, great men in the annals of Service history. It is invidious to pick out personalities but one or two must be mentioned. D A Boyle came with the September 1922 Entry. He was to become the first Cranwell cadet to rise to the position of Chief of the Air Staff and until 1978 was the Patron of the Old Cranwellian Association. In his entry was R L R Atcherley who was the first cadet to return to the College as Commandant. The list is compulsory reading for all who would understand what Cranwell has meant to the development of the Royal Air Force especially in supplying the nucleus and core of its regular officers.

The choice of officers to occupy senior posts at the College was made with great care at the Air Ministry. The first Commandant, Air Commodore C A H Longcroft, was a fine figure of a man. An officer of the Welch Regiment, he was trained at Sandhurst. He took the Royal Aero Club's certificate (No 192) in March 1912 and was seconded to the Royal Flying Corps. He distinguished himself by making a pre-war non-stop flight from Montrose to Portsmouth and back to Farnborough in 7 hours 20 minutes. Serving with No 2 Squadron in France at the outbreak of war, he became OC No 4 Squadron and eventually a distinguished Brigade Commander. A fine and keen rider to hounds, with handsome appearance, hat worn slightly angled, pipe in mouth, hunting crop or beagle whip in hand, he was the Commandant par excellence to the early courses of 1920-23.

The tone he set was echoed in the College Magazine and is borne out by reminiscences of former cadets. The Belvoir and Blankney Hunts met twice in the season at Cranwell and the Commandant and several officers attended the meet. One or two cadets who owned or borrowed horses were allowed to ride but usually they were too busy at their work. From the outset, however, there was a pack of beagles associated with Cranwell. Its first master was Squadron Leader C F A Portal and cadets were encouraged to support the healthy cross-country pursuit. Senior officers were happy to write in the Magazine about adventurous expeditions in far off imperial outposts or with squadrons in the Middle East. 'Sport on the Sutlej River, 1925' was no exotic memoir for those whose ambition was to join their seniors in India or Iraq.

Motor cycles were allowed but motor cars were not permitted to cadets on camp. Needless to say there were many powerful and strange contraptions kept at nearby garages such as Byard's Leap. It was aircraft which gripped the imagination and each year the cadets made the pilgrimage to the RAF Aerial Pageant at Hendon. Many former cadets were soon to be seen as performers and aces of the RAF's yearly shop window.

Flying was simpler, instruments were few and aids above the clouds or at night not available. Flying was dangerous therefore and though accidents were rare, the first death of a cadet, V O Gillmore, in a flying accident was recorded on 7 July 1922. A feature of Cranwell is the cemetery near the village church. From RNAS days through to the present there is a visible roll of those who have died in training or on duty at the nearby airfield.

Anecdotes of former cadets are personally significant. One or two are more general. Group Captain A H W J Cocks (1923-24 B Squadron) and Group Captain F E Nuttall (1921-22 A Squadron) both recalled with some sense of shock and trepidation the sight of an airman on Saturday morning Station Parade being dismissed the Service. The accused was marched on in snowy weather. The charge was read out. The Station Warrant Officer then removed the offender's Service cap, cut off the cap badge and replaced the cap back to front. He then cut off all the brass buttons, one by one, and the two shoulder 'birds'. The tunic was then removed and turned inside out and the airman was made to wear it back to front. He was then marched off parade.

Discipline was firm and punishment either by the cadets' disciplinary code or from officers and NCOs was condign and swift. The Station Warrant Officer was one Sergeant-Major Gorwood (formerly of the Lincolnshire Regiment). He like the Drill NCOs was a figure to be respected. He was usually one jump ahead. Sir Dermot Boyle recalled the morning when the cadets were astonished to hear a whisper that Cadet Bernard had become overnight, by the death of his relation, the Earl of Bandon. At the most public moment of the early morning parade Gorwood shouted out 'Flight Cadet Mr Lord Bandon, Sir, Do look up!' He had got the message!

It was an active life and there is little wonder that at the time of Princess Mary's (Viscountess Lascelles) visit to the RAF Hospital Cranwell on 24 April 1925, the following statistics were recorded about cadets in the term passing out:

Average gain in physical improvement	
Height	0·8 ins
Weight	6·5 lbs
Chest (normal)	1·3 ins
Chest (expanded)	1·8 ins

Progress, spiritual, mental and physical, was carefully monitored by Trenchard at the Air Ministry. He continued to take an active interest in the minutiae of College life. After his early visit in March 1920 with HRH Prince Albert, he was back on 20 December 1920 accompanying Mr Winston Churchill who as Secretary of State for War and Air inspected the College. His next formal visit to the College and the Boys'

Wing was in December 1922, when he brought Lady Trenchard with him. However, few weeks passed without some correspondence. On 2 November 1920 he was minuting the Director of Training and Organization.[2]

> I do not like all this reduction in the amount of work at Cranwell. I do not know what they will have left to teach soon. I have never found that an officer prepared much in his rooms in the past and I should like to know how much extra time to himself all this cutting out will give to a cadet — it must be a very large amount.
>
> I dislike the idea of omitting the 'Theory of Flight'. It can be touched upon very superficially but I am certain that a cadet must know broadly the reason why a machine flies.
>
> Secondly, I do not know why you want to increase the time for Science which does not seem to me to be so important as W/T. I do not know why you should omit instruction in bombing, they ought to know the rudiments of it.
>
> Lastly, I cannot agree to cutting out the Army and Navy organization, this must be taught broadly.

The syllabus was becoming already the bone of contention it seldom ceased to be throughout the life of the College. In his Commandant's Report for 1923 Air Commodore Amyas E Borton who had taken over from Longcroft in August of that year reported: 'There are still cadets entering the College possessing little or no scientific knowledge, and the method of incorporating them under the syllabus remains a serious difficulty.'

Finance also was playing its part. Trenchard reluctantly had to forego a scheme for encouraging the study of modern languages at Cranwell. An imaginative project to stimulate language study by provision of books, some prizes after proficiency tests and assisted travel to foreign countries was blocked by financial arguments that the expense of public funds on cadets' education was already great enough.[3] Financial worries were never far from Trenchard's attention. In February 1923 he wrote to Longcroft, 'I am just writing you a line privately to warn you that the Treasury are sending representatives down to Cranwell to go into, on the spot, the question of the increased establishment that we have recently put up in order that Cadets on leaving Cranwell may be fully qualified pilots.' Whether it was over this or a similar visit is not clear, but there was a very dignified reaction from Longcroft to the CAS about some implied deficiencies in his performance. 'I am deeply insulted by your remark that I had no tact in dealing with civilians.'[4] Both men were working under pressure.

Longcroft had been rebuffed by the Air Ministry in his desire to build a proper church at the College. At first church services were held in the Gymnasium which was located with the swimming pool in a hangar type building on the edge of the South airfield. The building was provided with a small alcove which served as a chancel and sanctuary and which could be concealed by a screen when the building was used as a gymnasium. This was the first Church of St Michael and All Angels at Cranwell. When the College was founded it was necessary to seek larger accommodaton. A hangar used as a workshop on the north side of the main road through the Camp was converted to become the Church. There were several unusual features about this Hangar Church. The font incorporated engine cylinder heads supported by parts of aircraft propellers. The lights were made from aircraft propellers. The College motto, 'Superna Petimus', hung in gold letters at the east end of the church. Also hanging were the banners of Nos 2, 3, 4, 5 and 6 Squadrons sent to Cranwell for safe keeping.[5]

The Rev Bernard W Keymer was appointed the first Church of England Chaplain for the College at the request of Air Commodore Longcroft in October 1919. His Presbyterian, Methodist and United Board colleague was the Rev A M McHardy and the Rev J T Skinner Law was the first Chaplain to the Boys' Wing. On Advent Sunday 27 November 1921 the Lord Bishop of Lincoln, the Right Reverend W S Swayne dedicated the Hangar Church. Thereafter it became the centre of Cranwell's Service and family worship and the focus for friendship with neighbouring Lincolnshire friends and families. Discussion groups were held on Sunday evenings after supper and visiting speakers in the early days included the Rev 'Tubby' Clayton of Toc H, Lord Hugh Cecil and Air Commodore Brooke-Popham. The church played a vital part in the life and training of Cranwell cadets.[6] The College prayer—that of St Ignatius Loyola—is itself an epitome of the virtues to be emulated by cadets brought up in the Christian faith:

Teach us
good Lord to serve
Thee as Thou deservest
To give — and not

1 The original RNAS group at The Lodge in autumn 1915
l to r: CPO Stevens: Paymaster L E C Tier: Lieutenant J H Lidderdale: Lieutenant J R Potter: Sub-Lieutenant F H M Maynard

2 Parade at Cranwell on 23 March 1920

3 First Inspection Parade, 20 December, 1920
W S Churchill: Air Commodore Longcroft: Air Marshal
Sir Hugh Trenchard

4 Avro 504 Ks of A Flight in 1923

5 Gravestone of an RNAS pilot in Cranwell village churchyard

6 Pulling down in 1979 one of the remaining RNAS huts on the North Airfield

7 Officers' and airmen's graves in the village churchyard, Cranwell

8 The Greigs' cottage at South Rauceby

9 Duke of York's House

10 The only notable reported accident on the Cranwell railway

11 The 'Cranwell Express' which ran between RAF Cranwell and Sleaford

12 The first view of Cranwell for entrants to the Boys' Wing, East Camp

HEARTY
GREETINGS
for
CHRISTMAS
and
the NEW YEAR.
1924-25
ROYAL
AIR
FORCE
CRANWELL.
~ LINCS. ~

14 After Church Parade on a wet Sunday morning

15 March past after Sunday Church Parade. Salute taken by H M The King of the Hellenes. Air Vice-Marshal F C Halahan and Flight Lieutenant R A George are on the left

16 RNAS Huts. Cadet lines

to count the cost
To Fight — and not
to heed the wounds
To Toil — and not
to seek for rest
To Labour and not to
ask for any reward
save that of knowing
that we do Thy will.

Discussions among the cadets led to one of the most significant events of the 1920s – the formation of the Old Cranwellian Association. It was founded at its inaugural meeting on 14 July 1926 with the Commandant, Air Commodore Borton, in the chair. The primary objects of the Association were laid down as follows:

a. To organize an annual re-union at the RAF College and to promote and organize games between Past and Present Cadets.

b. To found by means of a small annual subscription a fund to be used for the furtherance of College interests and charities.

Membership of the Association was originally open to 'Anyone who has been admitted to and passed through the RAF College' and also the Association had power 'to invite to become members or honorary members officials of the RAF'. This rule was amended later so that 'such honorary membership should not be conferred while officials held office at Cranwell'. In the beginning therefore the Association was very much in the hands of ex-cadets who were naturally very junior officers. Air Vice-Marshal C A H Longcroft was asked to become President and in addition to the serving Commandant, Group Captain L W B Rees, was invited to become a Vice-President of the Association.[7]

In the beginning the Association was interested in social, sporting and charitable activities. It gradually broadened its interests. It became the mainstay of the RAF College Magazine and contributed to several improvements achieved at the College, eg, the Squash Courts. Its very worthy efforts toward presentations to retiring servants of the College may be less well known. As the nature of the RAF College has become more complex in recent years so the problems of adjustment to new conditions have arisen for the Association and will be discussed later.

It would not be fitting to conclude a brief account of life at Cranwell in the 1920s without a reference to the aircraft flown and the Flying Training Wing personnel. The title of Chief Flying Instructor did not appear in the records until September 1930. From 1920-30 the control of flying training was exercised by those who held different posts at different times. The two officers commanding Flying Training Wing already noted were Squadron Leader C F A Portal and Wing Commander L W B Rees. When Rees became Assistant Commandant in 1923 he continued to control flying training of cadets. He handed over to Wing Commander E M Murray in 1924 and Murray handed over to Wing Commander D C S Evill in early 1929. There was thus considerable continuity in the pattern of training. The workhorse for training was the Avro 504K. This was a biplane powered by a 100 hp rotary Monosoupape engine of French design. It was a good training aeroplane despite its occasional throwing of castor oil and the odd cylinder at inappropriate moments.[8] The Avro 504N or Lynx-Avro with a re-designed undercarriage and fixed radial engine did not come in until 1927. Sopwith Snipes with a maximum speed of 121 mph were at Cranwell from 1921 to September 1927 and were used by instructors and for solo experience of senior flight cadets on advanced flying. There were Bristol Fighters, including the dual trainer, and DH9As. East Camp could boast of a Vimy for wireless training. The Snipe was replaced by the dual-controlled Siskin, another fighter type aircraft with a maximum speed of 150 mph. There were also less well-known aircraft which flew from Cranwell between the wars, notably those of the Long Range Flight. In 1927 Squadron Leader C R Carr and Flight Lieutenant L E M Gillman attempted to fly non-stop from Cranwell to Karachi. Their aircraft was an ordinary Hawker Horsley torpedo bomber. To give it the necessary range, it was fitted with extra petrol tanks. The first flight ended in a forced landing at night in the Persian Gulf. But the aircraft had already covered 3400 miles on a great circle route and had been airborne for 34½ hours. A second attempt by Carr with Squadron Leader Mackworth as second pilot ended in a forced landing at Martlesham and the third attempt with Flight Lieutenant Dearth finished up in the Danube, without serious casualty.

In the meantime Messrs Faireys Ltd of Hayes had designed and built the special experimental Long Range Monoplane. In April 1929

Squadron Leader A G Jones-Williams and Flight Lieutenant N H Jenkins flew in it non-stop from Cranwell to Karachi, a great circle distance of 3948 miles in 50¾ hours. In December 1929 choosing a new route which might yield following winds, the same pilots set out from Cranwell for S W Africa. They met stormy weather over the Mediterranean and ran into the foothills at the eastern extremity of the Atlas mountains. Both pilots were killed in the crash. This sad loss temporarily halted RAF experiments in long-range flying.

There was another flight from Cranwell which also ended in mystery and presumed disaster. On 13 March 1928 a jet black modified Stinson Detroiter with wires and struts picked out in gold trundled for nearly a mile westwards before becoming airborne. It was piloted by Captain W G 'Ray' Hinchliffe, a hero of RNAS days, and a noted civil airline pilot. His companion in the left-hand seat, because Hinchliffe had lost his left eye in an accident, was the Honourable Elsie Mackay, daughter of Lord Inchcape and a qualified pilot. They were attempting to cross the Atlantic from east to west to win substantial money prizes and to secure for Britain the Atlantic award. Elsie Mackay would have been the first woman to fly across the Atlantic. Sir Samuel Hoare the Air Minister had been persuaded to allow the take off from Cranwell whose runway was the best and longest known to Hinchliffe. But it was in the face of a strong north-westerly breeze that the aeroplane took off from the snow covered grass runway. Last seen at 1.30 pm over the extreme south-west corner of the Irish Coast, the plane *Endeavour* was apparently setting course for Newfoundland. Nothing was heard again of the two skilled and intrepid aviators.[9]

Sir Samuel Hoare's permission was given even more reluctantly to the presence at Cranwell of another figure of history—AC2 Shaw, T E, otherwise known as Colonel T E Lawrence or Lawrence of Arabia. Admitted for a second time to the Royal Air Force, Shaw reported to Cranwell on 24 August 1925 and was posted to B Flight (South airfield) and lived in Hut 105 in West Camp. Shaw wanted to be treated as an ordinary airman and to be left alone, particularly by the popular press, which had tracked him down and hounded him during his previous RAF service at Uxbridge and Farnborough. There are several testimonies to his behaviour at Cranwell. The Commandant, 'Biffy' Borton had known Lawrence in the desert and in Palestine. He was initially annoyed at finding Lawrence on his station without any prior notification. But he accepted him and unknown to others often invited Lawrence to the Lodge in the evening for talks. 'He did a lot for the library at Cranwell', said Borton, and of course there is a special collection of Lawrence memorabilia in the present Cranwell Library in College Hall. Shaw was another airman to Flight Sergeant Pugh of B Flight who wrote about him: 'It seemed his sole purpose was to be an airman of the lowest grade and rank and to be left alone with his Brough at B Flight Cranwell. He was hero-worshipped by all the flight for his never failing cheery disposition, ability to get all he could for their benefit, never complaining, and his generosity to all concerned till at times it appeared that he was doing too much for everyone and all were out to do their best for him'.[10] There is a sympathetic and sensitive appreciation of Lawrence by Rupert de la Bere in the College Journal of autumn 1935 shortly after Lawrence's death in a motor cycle accident. A fellow airman at Cranwell in 1925-26 was Mr Cyril Hobbs who had lately passed out from Halton as a Carpenter Rigger AC1. He remembered Shaw reading or writing in his hut while the others disported themselves in Lincoln or Newark. 'He was a good airman and a good comrade.'[11] Shaw was engaged on the abridgement of his *Seven Pillars of Wisdom* into *Revolt in the Desert.* He completed this at Cranwell with aid from two airmen friends. Wing Commander Sydney Smith, who had known Lawrence at the Cairo Conference in 1921 and his wife, Clare, arrived at Cranwell in late 1925 and renewed their friendship with him. But for the most part Shaw kept away from officers while he was at Cranwell. It was not surprising therefore that the General Committee Meeting of the Old Cranwellian Association on 19 March 1935 decided that 'Although A C Shaw had written the book [*Revolt in the Desert*] when he was stationed at the College, he had an insufficiently close connection with the College to warrant his being invited to become an honorary member of the Association.' The chances are that Shaw would have declined the offer had it been made. He volunteered for overseas service in India and left Cranwell in the autumn of 1926 after a stay of just over one year. His memories of Cranwell were happy ones. 'This is a very comfortable, peaceful, cleanly camp,' and again 'Technically it is so good and peaceful'. But as he wrote to Charlotte Shaw, the wife of George Bernard Shaw, 'Cranwell is not a good winter station.'[12] The heat and aridity of Miranshah were more to his liking at the end of 1926.

The latter years of the decade were on the whole days of hope and promise. The Treaty of Locarno and the plans for a Disarmament Conference at Geneva were pointers to an era of better international relations. But there were shadows being cast. In the Far East the

Japanese military were flexing their muscles. And the economic condition of England and the world generally was giving cause for concern. The General Strike of 1926 was past history and Stanley Baldwin was attempting to heal the deep divisions in British society which industrial disputes had exacerbated. On the cadets at Cranwell even the severe financial restrictions on the expenditure of the Armed Services which so frustrated their leaders in Whitehall made little impact. On the whole these were sunlit days. Flying, social pleasure and above all sport were happiness enough for young men. During the Autumn Term 1928 the Cadet College Rugby XV defeated both the RMC Sandhurst and the RMA Woolwich teams for the first time. On a more domestic occasion a remarkable cricket match at Cranwell between B Squadron who scored 238 and A Squadron ended with the latter only 12 runs short of victory. This was due to a determined and exciting innings of 194 by one cadet. His name was D R S Bader. Flight Cadet D R S Bader later became the famous fighter pilot, now Sir Douglas Bader CBE DSO DFC.

5 *The Boys' Wing 1918-1926*

The spotlight of public attention and Air Ministry planning had been focused on the RAF Cadet College in West Camp, Cranwell. But another aspect of life in the Royal Air Force never far from the planners' thoughts was apprentice training. Trenchard once said that flying and engineering were the basic requirements of the Royal Air Force. Skilled servicing and maintenance avoiding the errors learned in such costly fashion on the Western Front by the Royal Flying Corps—these were goals to be attained by young mechanics and fitters enlisted in the RAF's technical training schools. East Camp, Cranwell was to play its part in this endeavour.

Early days in East Camp

Both the Admiralty and the War Office had realized in 1917 that the armed forces needed more skilled tradesmen, especially in the air services. It was not possible to expand the RFC's School of Technical Training at Coley Park, Reading and a decision was taken to open a school of technical training for men and boys, separately, at Halton Camp. This would provide a new permanent home for those currently training at Reading or in the Polytechnic schools in London. Two other boys' training schools were proposed at Letchworth and Eastchurch but they became surplus to requirement after the end of the First World War. The School of Technical Training (Men) was established at Halton on 10 September 1917. The School of Technical Training (Boys) was formed as a separate unit on 9 October 1919 and received its permanent establishment on 1 January 1920. In March of that year it was named No 1 School of Technical Training (Boys).

However, pending the provision of permanent accommodation at Halton, the Air Council decided to use Cranwell for the training of boys.[1] A student population of 2000 with an instructional staff of 25 officers and 340 airmen was envisaged. A Wireless School for Boys, RNAS, had been opened in May 1916 and the magazine *Piloteer* referred continually to the W/T Boys and Boy Mechanics between 1916 and 1918. With the building of the new barrack blocks in East Camp (including hot water, real beds, mattresses and sheets for the boys) a Boy Mechanics School and then a Boys' Training Wing were formed. On 6 April 1918 the first contingent of RFC boys moved up from Halton to join their resident RNAS counterparts at Cranwell. They were to be trained as aeroengine fitters and in all the associated trades and skills—fitters, riggers, blacksmiths, coppersmiths, carpenters, turners and MT fitters and mechanics.

The amalgamation of the RNAS and the RFC to form the RAF was not accomplished without some expressions of dismay and regret. Novelty of equipment and the idiosyncrasies of instructors accentuated the problems of adjustment. 'We quickly became bi-lingual with a leaning towards the saltier side', wrote Wing Commander S E Townson,[2] 'Dining-hall became messdeck, cookhouse became galley but a trip to Sleaford on the little camp steam train "as going ashore in the liberty-boat" seemed a bit of exaggerated navalese'. The unit suffered from change and, as we have seen, Prince Albert was acutely conscious of the stresses and the transformation from a simple structure to a more complex unit.[3]

The unit, whose history is carefully indeed lovingly chronicled in the Boys' Wing Magazine 1922-26, had several changes of name. In March 1920 it became known as No 2 School of Technical Training (Boys) but in April-May 1921 it reverted to Boys' Wing, Cadet College, Cranwell and remained as such until in October 1925 its title became No 4 (Apprentices) Wing and as such it was transferred to Halton.

Among the adjustments made in the early days, RNAS personnel added 200,000 to their personal numbers. For example 'No 1234 Leading Mechanic A Killick substituted two stripes for his fouled-anchor arm badge and became No 201234 Cpl Killick, A in the RAF'.[4] This procedure paralleled the change in the numbering of RNAS squadrons to the 200 series.

Life in the Boys' Wing

Mention has been made already of the participation of the Boys' Wing in Colour Hoisting and Church Parades on the station. But from the outset West Camp and East Camp were almost two separate worlds. Yet the links were very real. For example, technical instruction took place at first in the established RNAS workshops in West Camp. Sporting contacts were also frequent but social contacts were virtually nil. In those days officers and airmen seldom met except on duty and paradoxically the airmen were extremely conscious of their status which had certain well-defined rights which officers were at pains to respect. But the Cadet College and the Boys' Wing had the same Commandant and through him the station achieved a unity. When the Boys' Wing Magazine published its first number in Spring 1922 it was Air Commodore C A H Longcroft who supplied the appropriate Foreword. '"Superna Petimus". If subscribers and contributors to the Boys' Wing Magazine will always keep this motto in view, they will soon develop into the type of man we require for the Royal Air Force and the success of the Magazine will be assured.' The Manager of the Magazine was the Rev. J T Skinner Law, the first Chaplain of the Boy Apprentices. He followed in the fine tradition of the Naval Padre, the Rev. J de Vitre, who from 1916-18 was the energetic and wise Editor and Censor of the RNAS's *Piloteer*.

It is astonishing to read of the full life and activities of the Boys. Their magazine, *mutatis mutandis*, could well be that of the cadets in West Camp. All the sports, societies, flying activities and enthusiasms are there. Their visitors were also distinguished. The Chief of the Air Staff was the Inspecting Officer for the first passing out term in December 1922. By this date the strength of the Wing was 1208 boys organized in three squadrons of about 400 boys and each containing a proportion of the four current entries. The entries were normally in January and September. Flying experience was available for the boys usually from the North airfield in their own unit aircraft. At this time they had a Vickers Vimy, a DH9A, two Avros and a Bristol with a Falcon III engine. The training in the Boys' Wing was geared to turning out skilled and intelligent tradesmen. But there was an attractive bonus for the best pupils. Cadetships were to be offered to the best boys which would enable them to join the next entry to the RAF Cadet College in West Camp. This was yet again the foresight of Trenchard to open the doors in the new Service to all talents. Two cadets, F S O'Hanlon and H E Nowell, joined the February 1923 College Entry and moved from East Camp to West Camp. When the apprentices eventually moved to Halton it was recorded that the

Cranwell Boys' Wing from 1920 to 1926 had produced 18 cadets, over 60 corporals and some 600 LACs. Three cadets ex-Boys' Wing had won College Prizes and H D Spreckley was the first aircraft apprentice to be awarded a Wakefield Scholarship. The boys could well ponder over the words of their Inspecting Officer, Field Marshal Sir William Robertson, on 30 July 1924. 'Some people say that when we go to war, we shall require no other people but airmen but I do not quite agree with that. We need industrious, hard working, patriotic and brave men. You have a great responsibility because without good mechanics, there can be no good Air Force.'

Among the last of the Boys' Wing to transfer as a cadet to the College was Frank Whittle (1926-28, B Sqn). He was to prove one of the most original and distinguished graduates of the College. Inventor of the British jet engine, his name is honoured in the building called after him that stands near to College Hall to the west.

It may be a surprise to know that the term 'Old Cranwellians' was used first by the Boys' Wing Association. In the Christmas 1923 edition of their magazine it is written: 'The Old Cranwellian Association is going well and will soon number the majority of the first three entries. Padre Stephens has made a great effort to organize an Old Cranwellians dinner in January next.' This was three years before the Cadet College organized an association of the same name.

The Boys' Wing owed much to their educational staff who were distinguished and talented. The first headmaster was Lieutenant-Colonel A F S Caldwell, a former infantry battalion commander, who moved to Halton in 1922. He was succeeded by H A Cox late of Oundle School. Members of staff who later had distinguished careers in the RAF Educational Service took a full part in all extra-curricular activities and helped to make the training of the RAF apprentice a valuable and much admired way of life.

To be posted to the Boys' Wing was not to be relegated to an unglamorous job. Many distinguished future leaders of the Royal Air Force served their time in East Camp. Flight Lieutenant the Hon R A Cochrane was No 3 Squadron commander in 1924-25 and among the adjutants of the Wing were Flight Lieutenant L N Hollinghurst and Flight Lieutenant H S P Walmsley. The Officer Commanding Boys' Wing R J F Barton was a most successful horseman and the winner of several point-to-points and it is somewhat remarkable that an apprentices magazine should include in 1925 a long account of the RAF Point-to-Point Races at Boothby Graffoe under the patronage of Group

Captain HRH the Duke of York. The Starter was the Earl of Londesborough and the Hon Sec of the organizing committee was—naturally—Wing Commander R J F Barton.

In the last issue of the Magazine, now called No 4 Apprentices Wing Magazine, the Air Officer Commanding RAF Cranwell, Air Commodore A E Borton, wrote in 1926: 'All the various and numerous activities and successes of the Unit have been duly chronicled in the pages of its Magazine which has proved a connecting link between all those who have passed out into the Service and the present members. But perhaps the most valuable work which the Magazine has done for the Wing has been to foster the traditions which have been built up, and the spirit of comradeship which has been established in the past four years.'[5]

Between 1926 and 1929 there was no separate boys or apprentices unit at Cranwell but in August-September 1929 the Electrical and Wireless School moved from Flower Down (now usually written as Flowerdown) to Cranwell to start a new era of technical training in East Camp.

6 *Other Ground Units at Cranwell before 1939*

The Electrical and Wireless School which moved into vacant accommodation mainly in East Camp in September 1929 had a distinguished tradition within the short span of RAF history. It began life in 1915 as a School for Wireless Operators in the Town Hall at South Farnborough.[1] In 1918 it became No 1 (T) Wireless School, RAF at Flower Down near Winchester and in July 1919 its name was changed to Electrical and Wireless School. The aircraft apprentice scheme was introduced there in 1922 and it is no small distinction to be numbered as a Flower Down 'brat'. By 1926, when accommodation became available, there were rumours of moves to Cranwell but the accommodation at Flower Down was very good and it was not until 1929 that reluctance and inertia were overcome and the School moved northwards. By this time the World War I pattern of short courses had given way to longer and more complete courses to suit the requirements of a regular force. In February 1928 Group Captain A L Godman became Commandant of the School. It was under his aegis that the move to Cranwell was made. For Godman it was a homecoming since he had been the first Assistant Commandant of the RAF Cadet College in 1920-21. The tendency of ex-Cranwellians to be posted back for tours in different capacities is one that is constantly found in the records.

In 1931 Officers' Long Signals Courses were planned for 14 months' duration and after the reorganization of the electrical and signals trades in 1932, Boy Entrants were introduced in 1934. They were to be trained in the trade of Wireless Operator. In October 1935 the School was split into two wings. No 1 Wing was concerned with training airmen as wireless operators. No 2 Wing comprised officers' courses, aircraft apprentice courses, boy entrants and refresher courses. By 1936 the School population was broken down as follows:

1103	airmen under training as W. Ops.
573	aircraft apprentices under training as W. Op. Mech.
575	boy entrants under training as W. Ops.
152	aircraft apprentices under training as Instrument Makers

The Chief Instructor at this time was Squadron Leader (later Air Vice-Marshal) J R Cassidy.

Passing out parades were held at the School in July and December and cadetships were awarded for the best performances. For example in July 1928, a General Duties cadetship was awarded to No 366190 LAA Shirley TUC who also won the Hyde-Thomson Memorial Prize.[2] In January 1936 the first entry of boy entrants passed out as Wireless Operators. The School was training Electricians, Clerks, Plotters and Telephonists at this time in addition to tradesmen in skills already mentioned above. The officer in charge of the RAF College Band, Flying Officer A E Sims made a signal contribution to the dignity of the Electrical and Wireless School when he dedicated a piece of original musical composition called 'Sparks' to Group Captain H Gordon-Dean, Officer Commanding the School on 11 April 1937.

The Signals Squadron of the School had its own aircraft which flew mainly from the North airfield. Before 1939 these were Wallace II, Valentia, Wapiti and Tiger Moth machines with the Magister just coming in.

By the outbreak of hostilities the range of technical training courses available at the Electrical and Wireless School was impressive. East Camp had something of the order of 2500 officers and men under training compared with West Camp's 150 or so. The courses provided included the following:

Officers (RAFVR) Signals	6 months
Aircraft Apprentices—W. Op Mech	2 years
Boy Entrants—W. Op	1 year
Airmen—W. Op Mech	32 weeks
Airmen—W. Mech	8 weeks
Air Gunners—W. Op	12 weeks

Airmen—Teleprinter Op	5 weeks
Recruits—W. Op	24 weeks
ACHs—R/T Op	6 weeks

The other main ground unit at pre-war Cranwell was the School of Store Accounting and Store Keeping.[3] This small unit had, like the Electrical and Wireless School, firm links with the early days of flying. The School was descended from the stores accounting units of the RNAS and RFC. After the RAF was formed the School was located with No 1 Stores Depot at Kidbrooke and it became a self-contained unit there in 1925. Its first Officer Commanding was Flying Officer T H Jolly. On 7 October 1927 a new Officer Commanding was appointed—Flying Officer A E F McCreary late of the RNAS and HMS *Daedalus*. McCreary stayed with the School for many years in one capacity or another. On 1 December 1932 the School moved to Cranwell and took over a building which was previously an annex of the Sergeants' Mess, West Camp. The building was converted into classrooms and offices for the small establishment. Flight Lieutenant E G Keeping was posted to command the School. Policy and training were the responsibility of the Cranwell Command and the School came under the Administrative Wing of the RAF College for administration and discipline.

During 1936 several changes took place. A scheme of pre-training for Storekeepers consisting of six weeks at selected units followed by six weeks at the School was introduced. Accommodation began to grow scarce at Cranwell—another recurrent historical phenomenon observable in the records. Flight Lieutenant A E F McCreary was therefore detached with the Officers' Course to No 1 Equipment Depot, Kidbrooke, to carry out training there. In December 1936 the name of the School was changed to The Equipment Training School (Airmen). At the outbreak of war the School with its exiguous establishment was geared to deal with 500 trainees at one time. Flight Lieutenant (later Squadron Leader) A E F McCreary was again appointed Officer Commanding as war with Germany was declared.

The other small ground unit at Cranwell was the Supplies Depot.[4] This was a junior officer's command and was mainly concerned with bread-making and distribution of foodstuff to a number of neighbouring RAF units. On the outbreak of war Flying Officer H J Bamber was in command.

7 *A Permanent Home for the College*

The sense of realism which Trenchard applied to the syllabus of instruction and other problems at Cranwell tempered also his ambitions for housing his officers at Cranwell and his apprentices at Halton in appropriate accommodation. The Air Estimates for 1920-21 contained an item of £300,000 for the total estimated cost of a new Cadet College building at Cranwell. But Winston Churchill who signed the estimates approved of only fifty pounds to be voted in the year 1920-21. The idea of a new building was early on under discussion at the Air Ministry.[1] On 18 February 1920 the President of the Royal Institute of British Architects sent to the Marquess of Londonderry, who was Under-Secretary of State for Air, a short list of four possible architects who could design new College buildings. But the Treasury were already taking a hand in the affair. They agreed with the First Commissioner of HM Office of Works that the design of a Cadet College could be carried out efficiently by officers of his department. But there were even greater bars to Air Ministry thinking and progress. There was a determined drive to cut public expenditure symbolized by the investigations of Sir Eric Geddes' Economy Committee. On 24 May 1921 the Chief of the Air Staff minuted his Secretary of State as follows:

> I have given the question of the Cranwell Cadet College very careful consideration. Originally it was considered essential that Halton and Cranwell should have permanent buildings of a suitable type for this new and growing Service, so that the Officers and NCOs, and men of the future should feel that they were properly housed and

that the Service was one they could look up to. Halton, as you know, has been started, as the temporary huts were really impossible to utilize much longer, and the number of boys being educated was very much greater. But Cranwell had a better type of huts, and they have been improved considerably as a temporary measure; and though they are far short of the type of accommodation provided at Sandhurst I think in view of the need for drastic economy next year and the year after that it will be perhaps better to suspend the consideration of building at Cranwell for a year, when the subject can be renewed again.

> If, purely from the point of view of economy, this course is accepted, it must be recognized that it will take two to three years, probably, after the order to proceed is given, to complete this College, and of course maintenance charges will be a bit higher.[2]

An Air Council letter dated 3 June 1921 was sent to the Director of Works at Storey's Gate to state that the Air Council was 'compelled to defer for the present any expenditure upon a permanent college. They hope, however, to be able to proceed with the work within the next two or three years.'[3]

At a time when the very Service was under attack from the Admiralty and the War Office, Trenchard knew what his priorities were. It was a case of *reculer pour mieux sauter*.

In the event it was to be several years before serious financial proposals for building would be acceptable to the Treasury. Plans for the new college had been drawn in the Office of Works but were put into storage. The subject kept coming up especially when the rising scale of maintenance charges on existing accommodation at Cranwell had to be faced. In 1924-25 there was discussion at the Air Ministry about moving the College to RAF Bircham Newton to acquire better accommodation. An even more unacceptable suggestion at least to Trenchard, was that cadets should be taught to fly in the aerodromes in the Canal Zone of Egypt. Both ideas were rejected. But a letter from Air Ministry to the Commandant at Cranwell on 17 April 1925[4] informed him that 'It must be recognized that the existing College buildings will continue to be used for at least the next three years.' Priorities for expenditure in the Royal Air Force were determined by the building programme of the Home Defence Force Expansion Scheme.

But the desire for a building to house its future officers worthily was never far from the minds of the planners or from that of the Commandant. On 14 June 1927 there was a first meeting of the Cranwell Cadet College Committee. Air Commodore Borton was in the chair with a representative from the Organization staff and an official from HM Office of Works. The Committee visited RMC Sandhurst in July to inspect the accommodation there and carried out a site inspection at Cranwell in the same month. At the end of the year an Air Council letter to the Treasury estimated the cost of a new College building at £260,000.[5] The estimate was based on plans adapted from those prepared in 1920[6] and assumed a cadet population of three squadrons of 45 each with the possibility of a fourth squadron to be added later. An interim establishment for the Cadet College was drawn up in July 1928 as follows:

Cadets	150
Officers	38
Officers to be accommodated in College	6
Civilian batmen for above	3
Civilian Education Officers	6
Civilian Staff	65
	268

The site as selected for the building was situated on the north side of the main public road passing through the station and immediately opposite the then Cadets' Mess. The building would necessitate the removal of the three large water towers also north of the public road. Submission of the building scheme was made to the Treasury on 19 September 1928 and approval in principle to proceed was given on 28 October 1928 to a revised estimate of £299,550. Following months of consultation and modification of drawings, Air Vice-Marshal F C Halahan, the Commandant at Cranwell, approved the building plans at an Air Ministry conference on 24 May 1929. It was an important conference because in addition to Halahan, who had his Assistant Commandant, Wing Commander D C S Evill, with him, there were present Air Commodore A E Borton and Air Chief Marshal Sir John Salmond and the architect from HM Office of Works, Mr J G West.[7]

The way was now reasonably clear for building to proceed. Messrs Bernard Pumfrey Ltd of Gainsborough received the contract for excavation of the foundations and Messrs J Chapman and Sons Ltd of

Leicester gained the contract for the erection of the superstructure.[8]

The scenery north of the main road underwent a transformation watched by cadets from their old hutted camp on the south of the road. Railway tracks were laid from the existing line to take building materials across what is now the Orange to the site of College Hall.

A ceremony took place on 29 April 1929 when Lady Maud Hoare, the wife of the Secretary of State for Air, laid the foundation stone of the new college.[9] It is known now that a clandestine, mock ceremony took place the preceding night at which cadets playing leading roles gave impressive caricature performances. The cadets clearly felt that the old order was changing.

The new Commandant, Air Vice-Marshal A M Longmore, had expressed his view that 'an unclimbable fence' such as had been erected around the old Mess to cut down certain irregularities by cadets, was not deemed suitable for the new building. Some thought was given at one stage to putting two gatekeeper's lodges at the main road entrance to the drive to the new College. But a more fundamental discussion took place over the lighting at the top of the building. The Commandant suggested to Colonel J F Turner from the Directorate of Works and Buildings that the dome could be floodlit and something more was needed than a red warning light for aircraft. One idea was to use sheets of Selenalyte glass illuminated from the interior and said to give a light visible two or three miles away. In the end the Air Council decided to install a revolving beacon light and Air Pilot (Volume I) carried the following amendment in 1933: 'An air light has been installed at the top of the tower of the Cadet College Building, 140 feet above ground level, at the Royal Air Force Station, Cranwell.' 'The beacon gives a white light with 20 flashes per minute', so Cranwell told the Air Ministry on 22 March 1933. 'The effective range of the main beam of light in fair weather is approximately 20 miles.' That warning beam has been a welcome sign of home and safety to many a worried traveller and to all neighbouring folk is a familiar cynosure.

The Journal described the new building in the issue of Autumn 1933 and drew attention to the avenues of lime trees given by Sir Samuel Hoare. It said the architecture externally and internally conformed to an English tradition. It was indeed a noble edifice. Pevsner's comment is somewhat patronizing: 'The style—a red brick stone-dressed Neo-Georgian Baroque—is what one would expect of the official grand manner of 1933... The plan consists of a roughly rectangular central block linked by narrow corridors to quadrangular blocks open on one side making a frontage of over 800 feet . . . A dome above of square substructure . . . the dome has all the usual appendages . . . pilastered drum, columns breaking forward and paired at the angles.'[10]

Within College Hall the cadets were to have living accommodation, lecture hall, lecture rooms, library, recreation rooms and laboratories—and the Mess. In front was a large circle of grass, the Orange, and a wide gravel frontage for drill and ceremonial parades. A collegiate atmosphere was achieved by the grouping of cadets' rooms around quadrangles as in the older British universities.

The Treasury had sanctioned expenditure to a total of £321,000 and the Air Ministry was well pleased with the accommodation it had achieved. A letter was sent to the Office of Works on 31 October 1933:

> On the completion of the new buildings at Cranwell and the impending transference of responsibility for their further maintenance by the Air Ministry, I am commanded by the Air Council to convey to the First Commissioner their warm appreciation of the work carried out by the Office of Works in connection with the construction of the College.

The letter was signed by Christopher Bullock, namely Sir Christopher Bullock, Secretary of the Air Ministry.

Before he handed over to his successor Air Commodore (Air Vice-Marshal 1930) A M Longmore, the Commandant Air Vice-Marshal F C Halahan saw the introduction of two new features of the College. First on 1 February 1929 the title of the College was changed to the simple and dignified one of Royal Air Force College, Cranwell. Second, the College on 19 December 1929 received from Garter, Clarenceux and Norroy, Kings of Arms, the grant of Arms of the College. These are described in admirable detail in two articles in the College Journal.[11] It will suffice therefore briefly to indicate the historical appropriateness of the Arms. They are based on the arms of the de Cranewell family discovered in the village church in the seventeenth century. They feature three cranes, emblematic of long-distance flight, with wings spread, signifying the three former squadrons of the College. The ground of the shield is light blue and the royal status of the College is embodied in the three lions in red and gold. The crest is fittingly the figure of Daedalus, the mythical first aviator whose name was borne by the parent ship of the RNAS stations including Cranwell. The legend *Superna Petimus* (We seek things that are above) had long been the motto of the College and is attributed to the Reverend B W Keymer, one of the first College chaplains.

To mark the 50th anniversary of the founding of the RAF College, Garter King of Arms and Windsor Herald approved in 1970 the addition of supporters, in the shape of silver eagles, to the arms of the College. Cranwell became the first Service establishment, it is believed, to have a full achievement of arms.

But Longmore had been replaced by Air Vice-Marshal W G S Mitchell before the official opening of the new College took place on Thursday 11 October 1934. HRH The Prince of Wales performed the ceremony and it was a glittering occasion. The Lord Lieutenant of Lincolnshire, the Rt Hon The Earl of Yarborough from Brocklesby Park, was the head of the reception party. With him was Lord Londonderry, the Secretary of State for Air and Sir Edward Ellington who was Chief of the Air Staff, accompanied by other Air Council members. From the dais of the main lecture hall His Royal Highness declared the College open. The Lord Bishop of Lincoln then offered the following prayer:

> Almighty God, under the shadow of whose wings the children of men do put their trust and whose everlasting arms are ever underneath us, we pray thee for thy blessing on the Royal Air Force and particularly on this College; upon all who shall bear command, and all who shall be trained, upon all who teach and all who learn therein, that in loyalty, discipline, duty and self-sacrifice they who go out from here may faithfully serve their King and country, and, in serving them, serve Thee; and as in this life they climb the steep ascent of Heaven, they may so give themselves in life and death, that they may in the end come to the full vision of Thee in thy service in the Heaven, eternal and unseen that is thy throne, through him who came among us as one that serveth and gave his life for us, thy Son, Jesus Christ our Lord. Amen.

The Prince of Wales remained to lunch with the cadets and the distinguished guests. There were 275 guests in the Main Dining Hall and the total bill it is very satisfactory to recall came to only £153 11s 11d, of which the College was asked to pay £46 1s 7d. Among the guests were many old friends of the College—General and Mrs Adlercron of Culverthorpe Hall, Commander and Mrs Amcotts of Rauceby, Lord Barnby of Scopwick House, Lord and Lady Brownlow of Belton House,

Lord and Lady Garvagh of Ashby Hall, Digby, the Earl of Londesborough of Blankney Hall, Lord and Lady Monson of Borton Hall, the Earl and Countess of Liverpool of Hartsholme Hall, Skillingthorpe, Lord and Lady George Seymour of Wellingore, Rear-Admiral and Mrs Lambert of Aubourn Hall, Lincoln and Colonel and Mrs King Fane of Fulbeck Hall. These and many others too numerous to list were present as representing the landowners and people of Lincolnshire amongst whom the new College building had come to stay.

8 *RAF College Life in the 1930s*

The decade of the 1930s leading up to the outbreak of the Second World War saw the College consolidating its pattern of training and way of life. The decade began confidently with the formation of a third squadron, C Squadron, in the summer of 1930. A Squadron moved into the brick huts to the west of the Cadets Mess while C Squadron occupied the iron huts on the east of B Squadron. The accommodation remained roughly the same as that of the 1920s entries. But across the road cadets saw with their own eyes the construction and drainage work that was to be the foundations of the new College building.

The number of entries remained at two per year but while the planning figures allowed for an entry of 50 cadets, the eventual passing out parade usually mustered about thirty on the senior term. In addition to students from the older Dominions, cadets from India began to appear regularly in the entries. The pattern of entry and recruitment remained the same, the majority of cadets coming from the public schools and grammar schools via the Civil Service Commission examination.

The development of Cranwell might well have been different if it could have relied on obtaining from more schools boys who were not only enthusiastic about flying and aircraft but with some knowledge of engineering practices. An innovation at Highgate School was a noteworthy experiment. In 1929 Highgate School acquired a new, very large and well equipped science building. The Headmaster, Dr J A H Johnston, a clergyman, was a mathematician of progressive outlook. With the help of his science and mathematics staff he developed an aero class to give practical training to those boys intending to go to

Cranwell or into the aircraft industry.[1]

A former member of the Royal Flying Corps, one Sergeant-Major Wilkinson, became a member of staff. He had workshops for wood and metal work and the Royal Air Force gave the school a Snipe and an Avro 504K, which were housed in a large hangar on the roof of the new science block. Gifts from aircraft firms supplied engines such as the Armstrong-Siddeley Jaguar, a Napier Lion and a Monosoupape for practical work. Theoretical instruction and demonstration usually preceded practical work over the whole range of aircraft maintenance.

There was also a regular complement of ex-apprentices awarded cadetships at Halton or from the newly established Electrical and Wireless School at Cranwell (see Chapter 6). In 1934 an experiment was tried when seven pilot officers who were university graduates arrived at Cranwell to complete their General Duties training. This was a short-lived experiment and the scheme was discontinued when the last of the ex-university pilots finished the course in March 1936.[2] The scheme was of course revived in a drastically altered form in the 1970s, ie, the Graduate Entry Scheme.

In September 1933 cadets took up residence in the still uncompleted new College building. The old Mess was vacated and was used for various purposes. A Roman Catholic Chapel was made in the east wing and other rooms were used for stores. In 1940 the WAAF had dormitories in the Mess and after the war, from 1945 to 1952, it became the Sergeants Mess and so it remained until it was demolished, not without regret and some ceremony, in 1953.

The daily life of the cadets changed very little. Flying was still the main activity and the general impression gained from the majority of former cadets is that the academic subjects were tolerated or even enjoyed in proportion to their obvious relevance to a flying career in the Royal Air Force. Some of the academic tutors were appreciated more for their personal qualities than the information they were imparting. Drill and physical recreation continued to be of greater priority than book learning. Outdoor pursuits were still encouraged by the example of the Commandant and senior officers. Almost all were enthusiastic horsemen and some were supporters of the beagles. The Blankney and Belvoir continued to meet annually at the College and the Hunt Club encouraged the younger officers and cadets to ride. Flight Lieutenant H C Pyper, a one-armed officer, was the great enthusiast for beagling. But when he left the College, support fell away and in 1936 the Cranwell Beagles were disbanded. The pack was sold and it was not until 1951 that a new pack, known as the Per Ardua (RAF) Beagles, was formed by Air Commodore L G Levis.[3]

Through hunting, beagling and rough shooting the cadets met some of the neighbouring landowners and farmers. Tennis parties, tea parties and dances were social events much prized by the cadets and the young ladies living near Cranwell. Apart from Saturday and Sunday the cadets still dined in Mess every night. Wednesday night in the early 1930s was the formal guest night. It was after the dinner that the Junior Term were exposed to the attentions of the Senior Term. The ordeals changed and being tossed to a very great height in a blanket began to be the most that 'sprogs' had to endure. The initiation period ended when a cadet's bowler hat was symbolically and actually destroyed by jumping on it. Then the cadet could call himself a uniformed officer under training.

The codes of conduct and discipline developed in the 1920s continued to be supervised by the cadets themselves. But mention must be made and credit must be given also to the guidance and almost paternal solicitude of some of the College civilian staff, particularly the batmen. When a cadet was called 'Filthy' on parade, the offence was as much felt by his batman. These civilians were also sources of help and advice to the uninitiated and the unwary. Some became almost legendary figures. There was Mr J Lewistein who served 29 years as a batman with B Squadron (he retired in 1959). He had a recipe for producing on boots and shoes a brilliance such as was not seen before on land or sea. Mr F Green also of B Squadron served for 44 years. He was the last to retire (in 1964) of the 11 batmen who came from Sandhurst to Cranwell in 1920. The Head Stewards were figures of great dignity. First Mr E Curt, who originally came to Cranwell as a naval steward in 1916 and did not retire until 1963. Then Mr F J Brown, who came to Cranwell in 1935. The present Chief Steward is Mr J W Collishaw, whose father and brother also served at the College. Mr A Clay MM, the Head Porter until he retired in 1959, joined Cranwell in 1923. He used to work for the farmer Mr Usher Banks at Lodge Farm before World War I. The Cadets' Mess Secretaries were also men of character and influence. Major A J Dawes and his family were remembered with affection by most cadets. In 1931 Wing Commander E J Sayer MC took over his duties. Few had the unbroken record of Mr J H S Lager MBE who was Head Clerk from 1920-27. He became College Accountant in 1927 and when the war came made a smooth sideways transfer to Station Accountant Officer from 1939 to 1947.

Motor cycles and later three wheeler cars were officially permitted, being within the limited pocket of the cadet. On such vehicles Nottingham, Newark and Grantham were accessible. Sleaford and Lincoln continued to be out of bounds. In the Air Estimates 1936 the establishment of cadets was given as 144 with an average complement of 133. The pay of cadets was 6s 6d per day and their parents had still to find fees of £100 plus another £100 for uniform and books. King's Cadetships were valuable and Prize Cadetships awarded on the results of examinations assisted with a reduction of college fees and free provision of uniform and books. It was an exacting life and a full life, by no means luxurious.

Flying had been under the control of the Assistant Commandant from 1923 until September 1930 when Squadron Leader M Thomas DFC AFC was appointed Chief Flying Instructor (CFI). He was followed by Squadron Leader C N Lowe who became a Wing Commander in 1933 and filled the post until March of that year. Cranwell had been the Headquarters of Red Force in the annual bomber/fighter exercises of that year. The AW Atlas had replaced the Bristol Fighter and DH9A as an army co-operation aircraft and soon the 504N gave way to the attractive Avro Tutor. The Hart arrived in 1933 and there were one or two Fairey Fox aircraft available for instructors and cadets capable of advanced flying practice. The Siskin was replaced by the Bristol Bulldog in November 1933 and the Audax replaced the Atlas in 1934. The basic trainer therefore was the Tutor. Cadets were assigned to A, B or C Flights whose lettering bore no relation to their squadron allegiances. In addition D Flight was a fighter flight or advanced training flight. The distinctive colours of wheels and spinners, red, yellow, blue and green were preserved.

Instrument flying was still rudimentary only and cross-country flights to Duxford, Bircham Newton or Catterick were mainly a matter of map reading and close observation of ground features. But the cadets were getting into the air much more quickly than in the 1920s and this made for happier students. There was much to be excited about. Not only were former cadets giving exhilarating displays at Hendon and other air pageants. Two former cadets, H R D Waghorn[4] and R L R Atcherley had won international fame by their activities, especially in the Schneider Trophy races and in the RAF's High Speed Flight. It was from their own South airfield that cadets saw Squadron Leader O R Gayford and Flight Lieutenant G E Nicholetts[5] take off in a specially equipped Fairey Monoplane on 6 February 1933 at the start of their record-making flight to Walvis Bay—a distance of 5340 miles in 37 hours 25 minutes. Cranwell had the great advantage of a very long runway for an east to west take off and the RAF's Long Range Flight made use of this facility from 1927 until 1938 when a flight of four Vickers Wellesleys flew non-stop from Cranwell to the Persian Gulf and back to Ismailia, Egypt, covering a distance of 4300 miles in 32 hours.

The airfield saw also its sombre moments. On 24 November 1931 three men working on land adjacent to the airfield rescued the pilot of a burning aircraft.[6] Less fortunate was Squadron Leader G H Martingell AFC who had replaced C N Lowe as CFI. On 20 February 1936 Martingell was killed in a flying accident. His successor was Squadron Leader (later Wing Commander) H T Lydford, who in turn was replaced on 19 July 1937 by Squadron Leader D A Boyle.

In October 1936 an Advanced Training Squadron was formed in addition to the Flying Training Squadron to bring the Cranwell cadets up to the same standard of flying training as that achieved at other Service Flying Training Schools. This note of seriousness was complemented by the appointment of Squadron Leader E J Kingston-McCloughrey, DSO, DFC as Chief Ground Instructor in March 1937 to direct and correlate the ground studies of cadets. Indeed the somewhat easy going tempo of studies was being looked at very critically and in 1938, following an inspection by His Majesty's Inspectors of Schools asked for by the Commandant, the academic syllabus was critically examined and several staff replacements made.

It was about this time, October 1937, that a mission from the Luftwaffe visited the College. This mission, coming from a Germany openly re-arming for war and in alliance with a Fascist dictatorship in Italy that had already caused alarm to our Navy and Air Force in the Mediterranean in 1935, was viewed with strong interest by the cadets. General Milch, Generalleutnant Stumpff and Generalmajor Udet were the principal visitors who were entertained in College Hall in October 1937. Following their visit Generalleutnant Wenniger, the German Air Attaché in London, visited the College in May 1938 to present portraits of Baron von Richthofen and Captain Oswald Boelcke, air aces of the 1914-18 War. Milch's secretary wrote on that occasion:

> On behalf of General Milch I am sending today a parcel containing a picture of Richthofen as well as one of Boelcke, to be used at

> Cranwell. I hope you like the pictures and approve of the idea of General Milch that they might encourage the feeling of mutual respect and help to prevent that our two Forces have to fight each other again as they were unfortunately forced to do some twenty years ago.

The College Journal[7] reported in dead-pan fashion—'The College is responding to this gift with pictures of Captain Albert Ball and Major Mick Mannock.' But the German visitors were smart, correct and impressive at least to one cadet then in his first term.[8] Much more so, in his view, than a French mission consisting of Général de Division J Vuillemin, Général de Brigade B D'Harcourt and the French Air Attaché in London, Lieutenant-Colonel P Fournier. They arrived at Cranwell in June 1938 and from them also in November 1938 came the gift of a portrait of the great French aviator Capitaine G M L J Guynemer.

As the decade wore on the possibility of war began to obtrude itself into the otherwise full and conventional life of the cadets. Air firing camps and armament practice camps assumed a new significance at Penrhos and Sutton Bridge. The lunchtime news on the radio generally produced an unaccustomed hush in the ante-rooms. In July 1938 the first Airspeed Oxfords arrived for twin-engined training. There were rumours of the new heavy bombers in production to replace the obsolete biplane bombers of the Heyford vintage and the monoplane Harrow. The Assistant Air Attaché from the British Embassy in Berlin, Flight Lieutenant W E Coope, visited units of the Royal Air Force to give a realistic and somewhat grim picture of the rearmed Luftwaffe. In 1938 and 1939 there were three entries to Cranwell, in January, April and September. Time for preparation for war was rapidly running out. The programme was reorganized into six instead of four terms with three passing out inspections in April, July and December.

As part of the gigantic work of reorganization, re-equipment and modernization of the Royal Air Force, a system of functional commands was instituted in 1936. On 9 July 1936 RAF Cranwell ceased to exist as an independent command and became a formation with Group status within Training Command. On 10 July 1936 a station headquarters was formed with Wing Commander J H O Jones as the first Station Commander. Then on 1 December 1938 RAF Cranwell became No 21 (Training) Group. The Air Officer Commanding was also to be the Commandant of the RAF College. At that time it was Air

Vice-Marshal J E Baldwin. A station headquarters was given the task of administering all the units at Cranwell. The College had indeed seen many fairly rapid changes[9] of Commandant and Assistant Commandant as the rearmament programmes post-1934 began to alter the shape of the Royal Air Force.

Those cadets who entered the College in January and April 1939 suffered the misfortune not only of having a truncated College course but also of having a concentrated flying training. Most were passed on to an Operational Training Unit by March 1939 and the September 1939 Entry of course had no training as cadets in the accepted sense of the word. The last Inspection took place in July 1939 when the Sword of Honour was won by T N Stack[10] who took the place of Bobby Morgan, reckoned by all to have been a natural for Chief of the Air Staff in due time. He was killed in an unfortunate mid-air collision.

As the war clouds gathered the cadets were heartened to know that their CFI had been awarded an AFC. A few months later he was on his way to HQ Advanced Air Striking Force with the British Expeditionary Force in France. Before he left England Wing Commander D A Boyle wrote a letter to all members of the Old Cranwellian Association recommending that the Association should be kept alive and subscriptions paid against the time of their next reunion. Of the replies he received 98 per cent agreed to this and also to the payment of a further sum to the RAF College Staff Pensions Fund. Very few anticipated that the next General Meeting would not take place until 13 June 1945.

9 *The War Years*

Instructional staff and cadets were recalled from leave to the College on 30 August 1939 in view of the grave international situation. Almost immediately after the declaration of war against Germany, the College ceased to function as such and was formed into a Flying Training School with an Intermediate Training Squadron and an Advanced Training Squadron. RAF Cranwell was to become a vast training base for a variety of flying and ground units much as it had been during the First World War. Both West Camp and East Camp were filled to capacity; both North and South airfields were used to congestion; and a variety of satellites and relief landing grounds were called into being to support the great training programmes being carried out. So diverse were the activities of the units at Cranwell that they must be considered separately. One thread runs throughout the war period. The existence of the College building and its facilities were carefully regarded and even in the direst moments, the thoughts of many at Cranwell were turned to a resumption of normality when peace should come. Many of the College civilian staff donned uniform but remained on the Station. Many ex-cadets and former officers returned in other capacities for duty during the war years. Cranwell never really forgot that it was the home of the College, despite the Babel chatter of tongues as officers and airmen from many nations passed through its corridors.

The pattern of organization of units in the first year of the war became rather complicated and it is as well to set it out. The senior formation at Cranwell was No 21 Group[1] whose Air Officer Commanding was also Commandant of the RAF College.

In February 1940 a new Signals Group, No 26 Group, formed at Cranwell before moving its headquarters to Buckinghamshire. On 7 July 1940 RAF Cranwell with its technical training and ground units was transferred to No 26 Group in Technical Training Command while HQ No 21 Group and the RAF College Service Flying Training School became lodger units on RAF Cranwell. Their parent command of course was now Flying Training Command since May 1940. Finally on 26 May 1941 RAF Cranwell[2] and its ground units passed under the control of No 27 Group in Technical Training Command and so remained for the war years. It was not until 1 April 1946 that RAF Cranwell reverted to the control of Flying Training Command. There were many other minor organizational changes involving units at Cranwell which will be noted as they took place.

The RAF College SFTS[3] as it was called, both suffered and derived benefit from having its superior formation on its doorstep. The rate of action was speedy. All flight cadets remaining at Cranwell were enlisted as aircraftmen. Many moved on rapidly to armament training at West Freugh in company with pupils of No 1 Course of the new SFTS. From there many moved quickly to join army co-operation and fighter units. At Cranwell it was an appallingly bad winter and all instructors and pupils were granted special leave from 1 to 25 February 1940 because of the snow and ice-bound state of the airfield.

The last entry of flight cadets into the RAF College before war broke out were gazetted to commissions on 7 March 1940 and thereafter the SFTS settled down to the arduous repetitive grind of flying training courses. Some of the early pupils were RAFVR officers who had come from No 3 Initial Training Wing at Hastings without the intermediate stage of an Elementary Flying Training School. The pilot wastage rate was about 25 per cent at the Intermediate Training Squadron and 15 per cent at the Advanced stage. By the time of the fifth war course the mix was beginning to reflect war conditions. Twenty-two were LACs who had been due to enter Cranwell as flight cadets in September 1939, 16 were Australians and three were already Pilot Officers. As training became more effective and streamlined, so the wastage rate fell sharply. But other features of wartime began to appear. Deaths in flying accidents became a regular occurrence. A sinister development was the interference with pupils' night flying practice during air raid warnings; on 26 February 1941 a training aircraft was attacked by a German intruder while attempting to land from night flying at Barkston Heath. Cranwell was indeed a worthwhile target for German bombers and the airfield and its supporting relief landing grounds received many hits from explosive and incendiary bombs in the next two years 1940-42.

The composition of the courses at the SFTS became international. For example on 28 May 1942, 20 Turkish officers, 4 Free French, 14 RAF and 2 Army officers together with 11 Free French, 9 Belgian airmen and 1 LAC arrived for flying training. The cemetery in Cranwell Village added new foreign names to its gravestones. Iraqi and Iranian students appeared and missions from Brazil, Bolivia and China visited the College in 1943-44.

In March 1944 the name was changed to No 17 Service Flying Training School and pupils and instructors were moved out of the College building into the Bristol Wood Mess. RAF Caistor was used mainly for night flying and Wellingore and Coleby Grange were also used for flying practice and relief landings. An indication of the size and administrative complexity of the unit can be gained from the following strength figures as at 31 July 1944.

	Officers	*Airmen*	*Airwomen*
RAF	107	1061	
WAAF	1		163
Turkish	55		
Belgian	3		
Free French		3	
Iraq	4		
Iranian	3		
Polish		1	
Czech	1	3	
Indian		1	
Fleet Air Arm		2	
RAAF	7	7	
RNZAF	7	2	

On 1 May 1945 No 17 SFTS moved from Cranwell to RAF Spitalgate and remained there until its disbandment on 18 June 1947. No 19 Flying Training School[4] formed at Cranwell to continue the work of flying training at the base.

Both the Group and the College SFTS had attracted a wide variety of distinguished visitors during the war years. These included Lord

Trenchard and an escapee from a German prisoner of war camp—Wing Commander B E Embry. Royal visitors included the Duchess of Gloucester and the Duke of Kent. HM King George VI himself visited the station on 13 November 1941. HQ No 21 Group remained at Cranwell until 28 July 1944 when it moved across country to Spitalgate.

These were difficult days when Station Defence Exercises were very real necessities, when staff officers were ordered to carry revolvers at all times in case of attack from German parachutists or other invaders and intruders. The presence of members of the Royal Family was a tonic boost to morale. The country had very little to celebrate up to that time in feats of arms. Other royal visitors were HRH Duchess of Gloucester on 6 May 1943 when she opened a Services' Canteen in East Camp and HM King George VI when on 13 June 1945 he visited Cranwell on the occasion of the Jubilee of the Royal Air Force College. The Reception Party was a very distinguished one. The Secretary of State for Air was the Rt Hon Harold Macmillan MP, Sir Charles Portal was Chief of the Air Staff, the Viscount Trenchard was there and the two Air Officers Commanding-in-Chief of Flying Training-Sir Philip Babington-and Technical Training-Sir Arthur Barratt. Air Commodore W E G Bryant was the Station Commander. It is very pleasant to recall that the King met during his visit many of the long serving civilian members of the College staff who were presented to him. These included members of the College Band who had served at Cranwell for 15 years' or more and some personnel who had served at Cranwell during His Majesty's own time there in 1918.[5]

There were two flying training units which had a short life at Cranwell in this period. No 2 Flying Instructors' School formed at Cranwell on 10 September 1940 under Wing Commander L Darvall, MC. The officers were accommodated in Officers' Married Quarters and the airmen used some buildings belonging to the Electrical and Wireless School. Tutors and Oxfords were the chief aircraft used by the unit. One of the Oxfords collided in mid-air over the aerodrome with a Wallace from No 1 Signals School on 4 October 1940. All crew members were killed. The unit officers were soon accommodated in the College Mess but the concentration of flying and the pressure of accommodation requirements persuaded the authorities to move the unit, now renamed No 2 Central Flying School, to RAF Church Lawford in May 1941.[6]

The other flying unit No 3 (Coastal) Operational Training Unit[7] made a more lasting impression. It moved to Cranwell from Chivenor on 6 August 1941 under command of Wing Commander J B M Wallis. The unit flew Whitleys, Wellingtons, Ansons and used Lysanders for drogue towing. Night flying was carried out mainly from Skellingthorpe. It was on a night training flight on 18 March 1942 that Whitley aircraft 'P' crashed into the roof of the College. The crew of three were killed outright. Serious damage was caused to the west section of the main building which caught fire. Fire brigades from as far afield as Newark and Boston were called out to suppress the fire. Three airmen in the College were injured in the crash and subsequent blaze.

The flying routine of No 3 (C) OTU could not be called glamorous. It entailed long navigational exercises over the North Sea and to all parts of the waters around the British Isles. The work was fatiguing and dangerous though flying safety was improved by an HF/DF mobile station built mainly by No 3 (C) OTU personnel. One extract from what started as a normal routine day's work may give a flavour of the activities of this unit. The entry in the Operations Record Book is dated 28 May 1942.

> Whitley 'L' flown by an instructor Flt Lt Crawford with six crew was airborne 2040 hrs on Exercise N5. It reached turning point (Rockall) at 0420 hrs and about 0450 hrs the port engine developed Glycol leak at 2000 ft. Flying on one engine, the aircraft was 'fixed' by DF station at Renfrew. The Captain was forced to ditch a few minutes later and the whole crew took to the dinghy. This was sighted about 0615 hrs by a Hudson aircraft sent on search. A destroyer was homed to the spot and the crew were picked up, transferred to an RAF launch off Greenock and taken to Abbotsinch where the Commanding Officer was waiting for them to fly them back to Cranwell.

For such excellent airmanship and determination as the above episode represents, the Air Officer Commanding No 17 Group, Coastal Command, wrote on 28 September 1942 'This total, [ie, night flying hours] the highest in Coastal Command, reflects the very greatest credit on all ranks and particularly those airmen engaged in maintenance.' Twenty-two courses passed through No 3 (C) OTU at Cranwell before the unit made its long journey south-westwards to its new base at Haverfordwest in June 1943.

The responsibilities of the Station Commander at Cranwell during the war years were onerous indeed. A list of the incumbents of that office is given at Annex IV. After some rather short tours of office in the early years, two officers Air Commodore H M Probyn and Air Commodore W E G Bryant held the appointment for the war years. The latter in fact died in Rauceby Hospital at the end of his tour of duty at Cranwell and was succeeded by Air Commodore R L R Atcherley on 12 September 1945.

In addition to looking after the innumerable royal and distinguished foreign visitors and the various missions examining training methods at Cranwell, the Station Commander had to cope with the often conflicting demands of the officers commanding the diverse units on the station. Accommodation and stores were always in short supply. There were occasions when the weather was so bad that airmen were marched eight deep back and forth into wind in order to flatten out the snowdrifts to allow aircraft to take off.[8] In July 1943 the western half of the South airfield was closed and handed over to contractors for restoration of the grass surface suffering from excessive usage. Formal inspections and surprise checks from at least two commands and at least three group headquarters could be expected. The mere task of housing, feeding and policing such a huge camp was very great. On the National Day of Prayer, 3 September 1942, there were about 7000 officers and airmen on parade. Again on the Battle of Britain parade on 26 September 1943 255 officers and 6310 airmen and airwomen marched past distinguished spectators like Lord Brownlow and Lady Londesborough. The numbers under training at a particular time are now seen to be quite staggering in view of the exiguous establishments of the modern Air Force.[9]

No doubt the Station Commander was highly involved in events of unusual significance. These could be periodic air raids on his station or its satellites. They could be defence exercises which were very much 'for real' in 1940 and 1941. They could also be the test flights on 15 May 1941 of E28/39 by P E G Sayers of the Gloster Aircraft Company or the F9/40 flown by M Daunt, airborne on 5 March 1943 from the South airfield. Jet flying had come to Cranwell. The Form 540 of RAF Caistor which operated as a satellite of Cranwell from June 1943 to June 1944 reveals how sedulously the Station Commander had to look after and visit his detachments, each with their accommodation, food, discipline and international incident problems. Cranwell indeed was during the war years as it has so often been since—the Royal Air Force in miniature.

If the flying units made the greatest demands on servicing and logistics, the ground units supplied the concomitant problems of handling large numbers of personnel, male, female and of mixed nationalities. A Recruits Training Wing formed on 1 April 1939 but rapidly moved away to Padgate on 30 May 1939. The Electrical and Wireless School under several names remained the major technical training unit at Cranwell. On 1 September 1940 the School changed its name from No 1 Electrical and Wireless School to No 1 Signals School. (No 2 E and W School had formed at Yatesbury on 1 December 1938). It had its first pupil casualty when L A C Fleming was killed in a flying accident on 4 October 1940. As the war continued so the School became the temporary home of officers and airmen of very many nationalities. The School's Operations Record Book shows representatives from the following countries: Canada, New Zealand, Newfoundland, South Africa, USA, Poland, Norway, Australia, Czechoslovakia and Free France. The routine work of the School was often subordinated to the urgent pressure of anti-invasion defence measures or fire-watching. But its leisure time activities were not neglected. The name of the School was changed again on 1 January 1943 to No 1 Radio School. At this time the unit could boast of:

No 1 RS Voluntary Band (30 instrumentalists)

No 1 RS WAAF Voluntary Drum and Fife Band

No 1 RS Apprentices Pipe and Drum Band.

Some WAAC personnel from the USA on Radio Telephony Operator courses enlivened the station in 1943.

There were many changes of command and a list of the officers commanding is given at Annex VIII. By the end of the war in the Far East, the School had achieved international recognition for its expertise in a wide variety of skills and trades. For example, it was natural to move to Cranwell for a time in 1943-44 the Beam Approach School technical training unit from Watchfield.

One of the many airmen who passed through the School was Corporal Godfrey Hounsfield. In 1979 he received the Nobel Prize for Physiology for the invention of the EMI body-scanner. His work on teaching aids at Cranwell came to the notice of Air Vice-Marshal J R Cassidy, AOC No 27 Group, who arranged for him to attend an electrical engineering college when he left the RAF in 1946.[10]

Two pieces of lighter commentary are culled from the Operations

Record Book. In 1942 it was reported: 'An astonishing arrangement exists in No 2 Wing Institute whereby separate queues are formed for tea and cake. Thus, after waiting five minutes in a queue for tea a man has to go to the tail of the next queue to get cake. Immediate action has been taken to remedy this hopeless state of affairs.'

Signallers are after all experts on circuit diagrams including short circuits. The second passage records[11] the Souvenir Concert which was given in the Station Cinema in February 1946 after the Final Report of the Allied Welfare and Entertainments Committee. A Polish airman who knew his Virgil improved the stage with the following silver lettering on a blue background: 'Haec Olim Meminisse Juvabit.' [The day may dawn when this will be sweet to remember.]

Some confusion is often caused by the numbering of the other major signals unit which existed at Cranwell in the war years. This unit whose site was to the west of the College started as No 1 Radio Operators School on 7 March 1941. Wing Commander W P G Pretty was the Officer Commanding at its formation. Its function was the training of Radio Mechanics (ground), Radio Operators (ground), both airmen and airwomen, and Filter Room Plotters (WAAF). It also ran courses for Signals (Radio) Officers. On 25 May 1942 the School was renamed No 1 RDF School and on 1 January 1943 it became No 8 Radio School. Like its sister unit in East Camp, No 8 RS received frequent visitors from Commonwealth and foreign services to examine instructional methods and to monitor the progress of their own nationals on courses. On 26 June 1944 the Operations Record Book[12] has the interesting comment: 'During the month, visits were paid by officers of operational squadrons to give talks to instructors and trainees in the interest of raising morale.' It was natural for HM King George VI to visit the School when he came to Cranwell on 13 June 1945. 'His Majesty saw examples of airborne radar in the Demonstration Room of the Senior Officers' Short Radar Course. He also saw CH and CHL equipments manned by crews of WAAF Radar Operators (Instructors) as well as a demonstration Filter Room also manned by WAAF Operators.' The School ended a distinguished wartime life under several officers commanding when it disbanded on 24 June 1946.

One ground unit which was deemed to be a worthier occupant of College Hall than the then resident Flying Training School was the Officers' Advanced Training School.[13] As the Royal Air Force expanded so the requirements for efficient administration and staff work also grew. The OATS formed on 1 March 1944 under direction of No

28 Group in Technical Training Command. The School consisted of the Senior Commanders' Course and the Junior Commanders' Course which had been previously conducted at the Junior Commanders' School at RAF Dallachy. The function of the Senior Course was to instruct potential station commanders in the principles of organization and administration of an RAF station. The function of the Junior Course was to instruct junior officers who, in the opinion of their Officers Commanding, were suitable for employment as Flight or Squadron Commanders, in leadership and discipline. A peak establishment of 200 students was set for this three weeks' junior course. The Commandant was Air Commodore S H Hardy and he remained with the School until he died in Rauceby Hospital on 8 April 1945. Shortly after this date an old Cranwellian, Air Commodore T G Pike, arrived to take over as Commandant and in July 1945 the unit now known as No 1 OATS moved across to Digby. It was at Digby on 19 October 1945 that the Air Member for Personnel, Air Marshal Sir John Slessor spoke to the staff and students on 'Future RAF Policy with regard to personnel' and gave his opinions on the strength of the post-war Air Force.

The School of Equipment Training[14] suffered more than most perhaps from lack of accommodation and from a very limited establishment of instructors in the early war years. Geared to an intake of seventy students per week in a fifty week year, the unit was under strain. By August 1940 it had a staff of only three administrative officers for a strength of 1300 airmen and airmen's married quarters had to be used to provide accommodation for recruits under training. Though the rank of the Officer Commanding was upgraded to wing commander in November 1940, plans were already being made to move the unit elsewhere.

WAAF equipment assistants were being trained separately at Melksham and since there was not likely to be sufficient suitable, accommodation at Cranwell the Equipment Training School (Airmen) was moved from Cranwell to Bridlington where a new School opened in May 1941 to train airmen and airwomen.

The School of Clerks Accounting[15] which ran officers' and airmen's accounts courses had an even shorter life at Cranwell. The unit moved up from Lympne on 31 May 1939 under command of a squadron leader. The unit was moved to Penarth on 28 January 1941.

The smallest unit at Cranwell was the prewar Supplies Depot.[16] This was a junior officer's command and at various times during the war the Depot came under the Officer i/c Supplies, West Riding Area of

Yorkshire, the North Midland Area, Nottingham or the Lincoln Area. During the severe winter of 1940 lorries could not get through the snow-bound roads and supplies of bread and meat had to be sent to outlying units by rail. About 200,000 lbs of bread were baked each month at the bakery at Cranwell and the feeding strength dependent on the Depot was about 15000 at its highest point. The unit continued to function through the war years until with reduced feeding strengths dependent on it and its bread ovens becoming unserviceable, the Supplies Depot was closed down at the end of 1949. This was part of the aftermath of war.

10 *Peacetime and some Difficult Years*

The months immediately following the end of hostilities were spent in run down and reorganization at Cranwell. The Educational and Vocational Training Scheme (EVT) got under way as hostilities-only officers and airmen/airwomen went back to civilian occupations. No 19 FTS instructors and students were still in occupation of the College building. It was a time for planning for the future. Indications of change came on 1 April 1946. RAF Cranwell was transferred from Technical to Flying Training Command. The Air Officer in command at Cranwell was to have the status of Air Officer Commanding a Group and would be Commandant of the RAF College when it re-opened. He would also command No 19 FTS. Thus the reins were placed in the strong and sure hands of Air Commodore R L R Atcherley. Air Ministry and Headquarters Flying Training Command decided that the RAF College would re-open in October 1946. The projected course was now to be for eight terms, about two and a half years in total. An intake of 40 cadets per term would give a peak population of 320 cadets. Six terms were to be spent in College with the status of flight cadets but two initial terms were to be spent in airmen's service. This was a political requirement and reflected wartime experience and thought. Fees were now to be paid by the Air Ministry. Life at post-war Cranwell was to show many changes from the pre-1939 pattern.

A syllabus working party was set up jointly by HQ FTC and the Director-General of Training to devise a syllabus of flying and ground training. The expert advice of the Empire Flying School was recruited in this study. A College syllabus committee was also set up with Wing Commander A C Kermode as its secretary to revise and streamline the ground training at Cranwell.

The first 15 cadets reported to the College in October 1946. They were not normal cadets. All had served in the RAF for short periods and most had done a university short course. The remainder of their entry were to join them in January 1947. They could hardly be expected to enjoy life in a barrack block as a squadron in the Apprentices Wing of No 1 Radio School in East Camp. Aircraft apprentices and a flight sergeant of the RAF Regiment gave them a rigorous initiation. Since there could be no automatic 'senior term' it was decided that 14 members of the senior course at No 19 FTS should be offered permanent commissions on condition they remained for a further year from April 1947 at Cranwell to provide the disciplinary and mess training for the junior term cadets. This entry from No 19 FTS was to be known as 45 Entry. Pre-war entries had used the expressions 'third' and 'fourth' terms to denote seniority. The system of numbering entries decided on in November 1946 thus began with a calculating error. Before the war in September 1939 there were only 43 entries to the College and this includes counting the first 1920 entry as two, ie, February (Naval) and February (School) entries. By April 1947 No 19 Flying Training School had departed for Feltwell and as the Operations Record Book noted with relief 'This leaves the College building clear for its proper purpose.' By April 1947 also, No 46 Entry had completed its two inital terms and was ready to move into the College. Flying training was not planned to start until a cadet's third term.

Early in 1946 the RAF Post-War Planning Committee had decided on the recruitment of a direct entry from civilian life to a permanent commission in the Equipment and Secretarial Branches. For this purpose a separate Wing was to be set up at RAF Digby as part of the RAF College. Group Captain R G Seymour CBE was posted as Assistant Commandant of the Digby Wing with Wing Commander A G Boycott[1] as his Director of Studies. Three entries of twenty-two per year for a course of eighteen months were planned. But applicants for the projected first entry in April 1947 were few and the Wing opened at Digby in autumn 1947. At that time there were 89 flight cadets in College Hall, 57 cadets in the so-called D Squadron in East Camp and 11 cadets at Digby. this was against a target figure ultimately of eight terms of 40 flight cadets and an Equipment and Secretarial Wing of 120. Among the innumerable problems facing the Commandant and higher authorities, recruitment was obviously a major anxiety. (In a discussion in the Air Council in 1946 the recruiting situation was diagnosed as 'the present phase of reaction against Service life.')

A conference of headmasters was arranged at the Air Ministry to ventilate the anxiety felt about the quality of entrants for the first three post-war courses. The reactions of some of the members of 46 Entry to conditions in East Camp could not have assisted in stirring up enthusiasm to join Cranwell. They had been treated previously as officer cadets at Perth and Hereford and for the first six months at Cranwell they had had no flying. Additionally there were shortages of fuel in the hard winter. Life seemed grim and unsatisfying.

But normality was being established. A Flying Wing under a Chief Flying Instructor formed on 18 April 1947 with A Flight on Harvards and B Flight on Tiger Moths. There was also a Passenger and Navigation Flight operating Ansons. The flight cadets in College Hall were organised into two squadrons, each with a squadron leader and a flight lieutenant.

Other problems were shortages of equipment and suitable laboratory and academic accommodation. The appointment of a Director of Studies on 1 July 1947 was a step forward. This was Mr Patrick Johnson, OBE, MA, of Magdalen College, Oxford. College extramural activities also were revived and a College Society was formed to finance and stimulate various indoor and outdoor recreational pursuits. Evening lectures by prominent speakers were organised after formal dining-in nights. No one would have been happier to know this than Rupert de la Bere who died on 4 January 1946 and was buried at his own wish in the Cranwell Village churchyard. By September 1947 the cadets were able to man three squadrons and gave a good account of themselves with a PT display and on parade with the band and trumpeters to Beat Retreat when Cranwell was At Home on 20 September 1947. But No 49 Entry of 32 Cadets had only 16 from the direct entry route. Eight were ex-apprentices and eight were former serving airmen. There were only seven instead of 22 applicants for the Equipment and Secretarial Wing at Digby. Thoughts were already turning to greatly improved long term accommodation for ground training. A new instructional block was proposed to be built to the west of College Hall. In the meantime refurbished huts on the former Radar 'A' site were made available. Science instruction was spread over nearly a mile at this period. There were shortages also of aircraft and flying instructors and the introduction of Prentice training aircraft exacerbated these deficiencies. It was a difficult time and the first internal dissensions arose between the claims of the ground instructors trying to follow a coherent syllabus and flying instructors who had to fly whenever aircraft and weather were available.

By summer 1948 D Squadron was disbanded and junior flights were moved from East Camp to a barrack block in West Camp though still remaining under airmen's conditions. The junior flights were also allocated to the three College squadrons. This move to closer association also coincided with the withdrawal from College Hall of almost all instructional classes except navigation. College Hall was becoming simply a domestic building. The highlight of 1948 was undoubtedly the visit of HM King George VI who on 6 July presented a Colour to the College, the first to be presented to a unit of the Royal Air Force.[2] It was also approved that one squadron in the future would be styled 'The King's Squadron' with the privilege of sitting at the centre table in the Dining Hall with the Colour unfurled in its stand on the Eagle behind the President's chair. The order of merit of passing out entries would determine the award of the style of King's Squadron.

In April 1948 Transport Command Dakotas had taken cadets for a memorable first liaison visit to the Ecole de l'Air at Salon. The first post-war passing out ball had been held and vacation visits to BAFO and to Navy and Army units were re-establishing a pattern familiar to pre-1939 flight cadets. But changes were still occurring. The Prentice arrived for flying in autumn 1948 and a Vampire jet aircraft was added to Flying Wing. In College Hall a collection of historical documents was received by the Library. This included Lord Dowding's famous letter to the Air Ministry in May 1940. Some time earlier, RAE Farnborough had given to the College a number of German aircraft, weapons and items of equipment which were meant to be the nucleus of a historical museum. Among his last important actions as Commandant, Air Commodore Atcherley appointed a committee to plan and advise on the hanging of all the College pictures. These included those from the Duke of Newcastle's collection, those on loan from national galleries, personalities of early days and Second World War portraits and pictures. When he handed over to his successor, Air Commodore G R Beamish, in February 1949, Dick Atcherley could feel sure and proud that the Cranwell he cherished had been re-established on sound lines.

Many problems beset the College in the post-war years but there is no validity in the argument that the College was not in the right hands. From 1946 to 1958 the Commandants at RAF Cranwell were Old Cranwellians, former cadets of the College. From 1947 to 1963 the

Assistant Commandant too was an old Cranwellian, a former cadet of the College. There could be some weight in an argument that the Commandant lacked the degree of autonomy given to the Commandants of the sister academies of Dartmouth and Sandhurst. HQ Flying Training Command and the Air Ministry separately or jointly monitored or controlled the policy and direction of the RAF College. To examine this question further is not germane to this history except to indicate the parts played by higher authority in the design of flying and ground training syllabuses and in connection with the pervasive factor of recruitment.

The decade opened with a ceremony that symbolized the heavy toll of war and was a fitting tribute in the Cranwell style. On 20 May 1951 the Cross of Sacrifice erected by the Imperial War Graves Commission in the Cranwell Village churchyard was dedicated by the Lord Bishop of Lincoln. The Cross was a memorial to the officers and airmen of RAF Cranwell who lost their lives in two World Wars and are buried in the churchyard. Air Chief Marshal Sir Arthur Longmore (Commandant, 1929-33) gave the address. There was a sense of continuity and shared tradition.

Sporting fixtures against old rivals were resumed. The traditional rivalries and exchanges of courtesies and visits between Greenwich, Sandhurst and Cranwell continued to rouse all the old enthusiasm and effort during the 1950s and early 1960s. Dartmouth was added to the list of opponents when the Royal Navy revised the age for entry to the Britannia Royal Naval College in 1955 and this provided suitably aged sportsmen at rugby, cricket and hockey. New opponents were the cadet teams from the RAF Technical College, Henlow.[3] New contacts were made — sporting, social and academic — with Oxford and Cambridge colleges and increasingly with departments of the nearer universities of Sheffield, Nottingham and University College, Loughborough. The cadets as always had a wide range of vacation activities: among those mentioned in the Operations Record Book were gliding, mountaineering, speleology and sailing. Visits to units of the three Services were frequent. For example, the cadets had been airlifted to see the operation of 'Plainfare', the Berlin Air Lift in Germany. Initial contacts were made with West Point and USAF establishments and with academies nearer home in France and Sweden. A happy liaison was established with officials and miners of Annesley and Newstead Collieries in Nottinghamshire. Reciprocal visits to see each others style of life and work were successful and educative. The local connection was also reinforced when the Chairman of the Sleaford Urban District Council (Councillor A Buttler) took the salute at a parade to commemorate the 35th anniversary of the opening of the College on 6 February 1955. Visits of headmasters and careers masters were numerous and much encouraged by the Air Ministry though demanding of staff time and College hospitality. Annual events such as the Ferris Drill Competition with its team of examiners from the Army or Royal Marines reinforced the periodic graduation parades of departing entries to reassure most people that Cranwell was getting back to its pre-war status. Some changes occurred. In 1953 the disciplinary structure of the flight cadet squadrons was changed. Corporal and sergeant ranks were abolished and each cadet squadron was given one senior under officer and three under officers.

Graduation parades at first in April, July and December of each year and from 1960 in July and December became works of art—studied and carefully rehearsed spectacles delighting the many distinguished visitors and the proud parents and friends who regarded them as a high point in the year's military and social calendar. The marching and ceremony seldom failed to stir the blood and emotions even in the often dire Lincolnshire weather.

Organizing and mounting a passing out parade involved long term and detailed planning. The choice of date and reviewing officer were critical decisions on which invitations to VIPs, parents and friends of the cadets, local dignitaries and service representatives had to be dependent. The form of the parade has changed with the years. Formed fours of course gave way to the new drill in threes. Four squadrons instead of three, or two, added complications. Other changes came with experience. Not least perhaps was the change which put the Passing Out Ball on the night after not the night before the Parade. Many a Reviewing Officer must have become aware in the olden days of odours of hop and grain not commonly experienced on the fresh Lincolnshire breezes.

The reception of the Lodge Party and the hosting of visitors, latterly in the usefully placed Whittle Hall were preludes to the assembly of guests on the seats alongside the dais facing the entrance of College Hall. The cadets formed up behind the East and West Wings of the College while the markers were placed in position and the College Band was poised by the main entrance. The squadrons marched onto their markers and were eventually handed over to the Parade Commander in line at the open order. The distinction of commanding the parade was given to the Sword of Honour winner. From the time it was given to the College, the King's and then the Queen's Colour was marched onto parade at the present arms.

The Reviewing Officer's car entered the main gates and moved round the Orange on the west side towards the saluting base. The flypast when arranged coincided with his arrival at the dais and the band played a fanfare of welcome. The General Salute, or on the royal occasions, the Royal Salute, was then played.

After inspection of the squadrons by the Reviewing Officer came the march past in slow and quick time. Then came the second General Salute and present arms after an advance in review order for 15 paces. All this done with the breath-holding precision and ease born of many hours of hard rehearsal in the early mornings of previous weeks.

The Reviewing Officer's address and/or presentation of wings and prizes usually followed, though the method and occasion for presentations have changed many times in recent years.

Perhaps the most moving moments of the parade came when after the Colour had been marched off, the graduating entry wheeled in pairs up the steps of College Hall and into the building to the strains of Auld Lang Syne from the band. No horse mounting the steps as at Sandhurst but cadets now officers moving off to a new, fuller life in the Royal Air Force.

Though shivering under blankets wrapped round the knees against the winter or spring blasts at Cranwell, neither parents nor friends could help but feel also a tingle of excitement and pride as their own particular cadets moved on. Even the hard-bitten staff who had seen it all before never became impervious to the stirring event.

HRH The Princess Elizabeth took the salute at the passing out parade of No 53 Entry on 1 August 1951 and the Royal connection was maintained when a contingent of flight cadets formed part of the route lining force in the area of Parliament Square at the Coronation of HM Queen Elizabeth II in June 1953. On 28 July 1953 HRH the Duke of Edinburgh was the Reviewing Officer at the graduation of No 59 Entry.

Cranwell was sufficiently interesting for the visit of a party of 12 London correspondents of foreign newspapers in May 1951. The Foreign Office had arranged the visit to demonstrate some of the activities of the Royal Air Force in preparing for the defence of the West. In September 1956 a party of senior Soviet Air Force officers including the Soviet Chief of Air Staff Marshal P Z Zhigarev toured the College. Somewhat proleptically on 14 June of the same year Air Vice-Marshal D F Barnett CBE DFC had given a lecture to the flight cadets in College Hall on 'The Influence of Air Power in the Middle East'. Not many months later Barnett was appointed as Air Commander of the Suez Task Force. Cranwell appeared to be back to its old form, in the thick of things and going well.

But in spite of appearances Cranwell was undergoing some stresses and strains. These derived largely from problems of control and responsibility for flying in general, aircraft movements and maintenance on the station. Until October 1952 another major unit, latterly No 6 Radio School, had shared the camp with the College. The Commandant had had the style of Air Officer Commanding RAF Cranwell. In place of the Radio School another unit formed at Cranwell on 15 November 1950 and remained there until 24 March 1953. This was No 3 Initial Training School[4] under the command of Group Captain L G Levis (1931-33: C Squadron). Its function was to provide officer and initial training for airmen selected for pilots and navigators. This was the time of urgent expansion of the Royal Air Force after the demobilization run down.

The Cold War, especially the threatening international situation over Berlin, and the shooting war in Korea were the background to this expansion. The unit moved into the huts vacated by the Radio School and a separate Officers' Mess was established in East Camp. The Aircrew Transit Unit was moved into Cranwell from Driffield. Eventually a separate wing of the ITS formed at Digby. This Digby Wing came under the command of Wing Commander M D Lyne AFC (1937-39: B Squadron) and was called No 3 (Digby) Wing of No 3 ITS Cranwell. Among its activities the ITS included the training of Saudi Arabian, Syrian and Burmese cadets and during the period of its existence it trained 1544 pilots, including 119 trained under the National Service Acts and 1088 navigators including 178 National Servicemen. It also used Airwork Ltd for a civilian-manned grading school. Thus for a time the AOC and Commandant had the responsibility for the efficient functioning of two flying training units on his station producing aircrew who were on sharply contrasted conditions of service. The rivalry particularly in flying efficiency was stimulating. The College was conscious of not being the only unit on the station.

The Operations Record Books (F540)[5] show that the AOC and Commandant made several changes in his pattern of organization.

Initially there was a group captain known as Officer Commanding Headquarters Wing. In September 1948 the Commandant's senior staff officer became the Group Captain Adminstrative, HQ RAF Cranwell. In September 1950 West Camp became known as the RAF College and the post of Group Captain Executive replaced that of Group Captain Administrative. During this period, however, the Assistant Commandant of the College through the Chief Flying Instructor exercised command of the personnel of Flying Wing (actually from March 1949). By 1956 Cranwell was free from what could be called 'other units' and it was reorganized as one unit in September of that year. From March 1957 onwards only one Operations Record Book return was made monthly. That was signed by the Unit Commander, RAF College Cranwell. In January 1960 the post was upgraded to group captain and on 11 January 1962 HQ Flying Training Command agreed that the unit commander should become the station commander but that on all official correspondence he remained 'OC RAF College Unit'.

There were three major problems which constantly occupied the attention of all concerned with Cranwell in the 1950s. These were accommodation; the syllabus; and the quality and quantity of flight cadet entrants. These problems need to be examined separately but their interaction was dynamic in its consequences during this period.

The RAF College Journal in June 1952 commented: 'A little has been done to relieve the straitened state of instructional accommodation. Five years ago plans were well advanced for a comprehensive instructional block in a style in keeping with that of the College but the plans have remained on paper.' The state of instructional accommodation continued to be unsatisfactory throughout the decade and was the subject of adverse comment both in the report of the Melvin Committee[6] and in the Report by HM Inspectors of Education[7]. Successive Commandants did what they could and were more successful in what might be called local improvements not wholly dependent on financial and logistic support from higher authority. In November 1950 the war scars of unbalanced iron gable and new brick pediment were removed and the roof of the west wing regained the symmetry it had before the Whitley crash of 1942. A scenic transformation was effected in 1951 when the entire system of hutted lines between the road and the permanent buildings south of it were removed. From the huts of the RNAS Camp only the Old College (now the Sergeants Mess) and the Post Office remained. The picture beginning to take shape was of the Senior year of cadets in the College, the Intermediate year in the transformed barrack blocks and the Junior cadets in the South lines behind them with a College Junior Mess established in the then NAAFI building of West Camp. By 1953 this Junior Mess was in full swing. On 8 September 1953 the Equipment and Secretarial Wing of the College moved from RAF Digby and 56 Cadets (39 Equipment and 17 Secretarial) were amalgamated into A, B and C Squadrons at Cranwell. D Squadron temporarily ceased to exist.

The College Building itself was enhanced by the present of a peal of bells from the Shell Petroleum Company and its associates. The bells were to play Retreat each evening as the Ensign was lowered. On 22 June 1952 the College Memorial Chapel was dedicated to commemorate the 477 ex-cadets, their names recorded in a Roll of Honour, who lost their lives out of a total of 1096 who were trained at Cranwell before the 1939-45 War. The Memorial Chapel, known as St. Michael's Chapel, was on the second floor on the south side of the west wing of the College. It was to be used for Holy Communion and private worship. Among the munificent bequests later transferred to the new Church were the altar and altar silver subscribed for by ex-cadets and by the relatives of those cadets who had lost their lives. The Commanders-in-Chief of the RAF Commands, the USAF and the Royal Navy made presentations of pews, font, Sanctuary Chair and Desk and the Sanctuary carpet (from Isfahan) was given by the Royal Pakistan Air Force.

In June 1953 the metamorphosis of West Camp was almost completed when the last remnants of the old Mess Decks—original College—Sergeants Mess—were pulled down and carted away. As the College Journal noted, 'The sole remnant of the naval corrugated iron perpendicular style of architecture remaining within the campus area is the Post Office where Mr Robinson commands a lonely ship in the midst of a grassy sea.' In 1960 the last relic was itself removed and the Post Office given a permanent site opposite the Guardroom in East Camp. The old familiar green hut disappeared. Mr J W Robinson the original postmaster of 1916 had died in 1955 but his son, Mr S Robinson, carried on the family name and tradition.

The advent of jet aircraft also meant that the runways had to be hardened and in 1954 the South airfield was handed over to contractors. During this period the Chipmunks operated from the North grass airfield and the Balliol squadrons flew from Barkston Heath. The aircraft sequence used in training post—1946 had been:

Tiger Moth—Harvard
Prentice—Harvard
Chipmunk—Harvard
Chipmunk—Balliol

A list of aircraft used in training through the years at Cranwell is given at Annex IX. In 1956 with the introduction of the Piston Provost-Vampire sequence of training, Nos 3 and 4 Squadrons flew from the South airfield and Nos 1 and 2 Squadrons operated from Barkston Heath.

Decisions such as the lengthening and hardening of the runways of the South airfield had their origin in Air Council discussions.[8] The question of awarding wings was involved. It was appreciated that the award of wings had become the natural and traditional accolade for aircrew at the conclusion of their cadet training. But the award of wings before completion of advanced flying on jet aircraft was not considered appropriate. Therefore it was necessary to include advanced flying in the Cranwell syllabus. There was indeed an alternative proposal to move the cadets away from Cranwell for the whole period of advanced flying or even to transport them daily by air to do their training at an AFS. Thus the syllabus, the length of the Cranwell course, the accommodation of the station and a real estate investment in hardened runways were all involved in an apparently straightforward decision to develop the runways at Cranwell and not at Barkston Heath. It is apparent from the records that more than one senior officer found it difficult to think of Cranwell as something very different from the normal flying training school and that flying, though the main priority, had to be linked with other considerations in the training of the flight cadet. This aspect of conflict between what the Hartley Report[9] diagnosed as 'the immediate' and 'long term objectives' of the Cranwell training will be discussed at a later stage in more detail.

While the domestic accommodation and the general appearance of West Camp were much improved by the mid 1950s, it was obvious that a considerable building programme was required to make the College an institution where boys, parents and educational bodies could believe that the Cranwell authorities took their technical and academic responsibilities seriously. At the direction of the Air Member for Supply and Organization, the Melvin Committee[10] was set up in May 1957 to assess the long term requirements of the RAF College in

17 **Fairey Monoplane with College in background**

18 The 'Endeavour' in which the Hon. Elsie Mackay and Captain W G R Hinchliffe took off from Cranwell, 1928

19 The Clock Tower above the Drill Shed to the east of the Parade Ground, West Camp

20 **Bristol Fighter**
21 Bristol Bulldog
22 Avro Tutor
23 **Hawker Fury I**
24 Fairey Fox
25 Armstrong Whitworth Atlas

K
5678

J
7948
J7948

26 Lady Maud Hoare laying the foundation stone of the new College building on 29 April 1929

27 The formal luncheon in College Hall after H M The Queen had presented the College with a new Colour, 1960

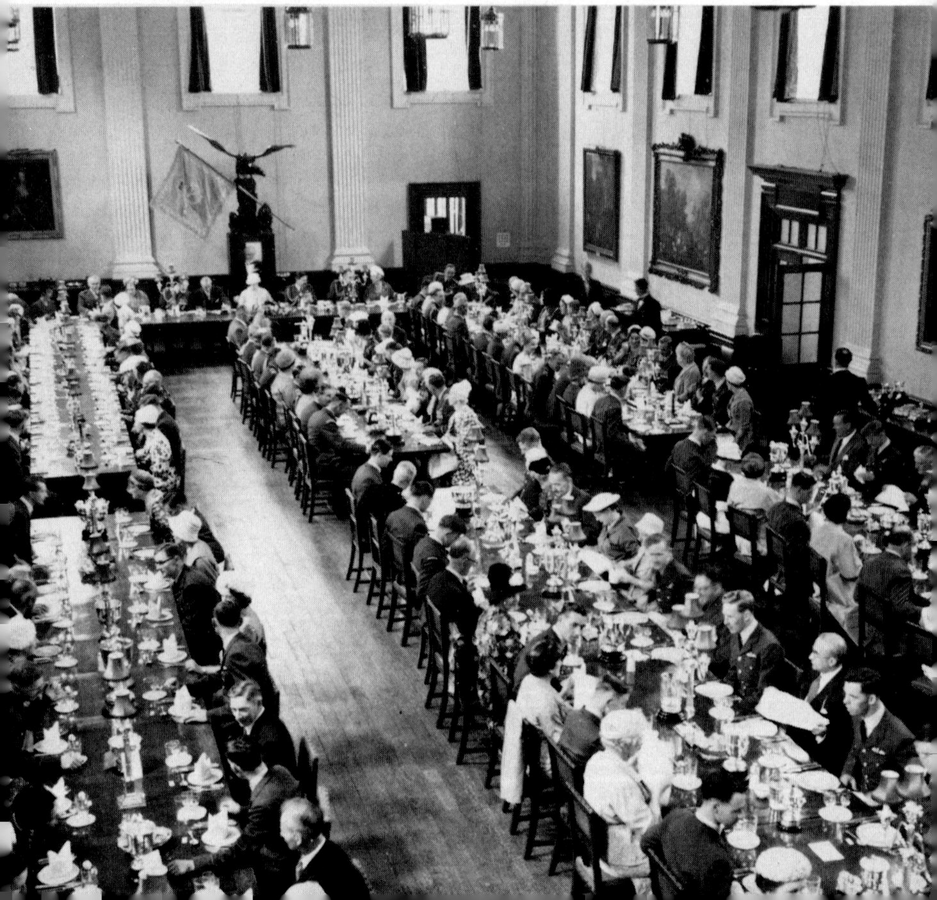

28 Presentation of the new Queen's Colour on 25 July 1960. Behind HM The Queen is the Commandant, Air Commodore D F Spotswood.

29 A passing-out parade

30 The Attestation

32 Interior of St Michael and All Angels Church

31 HRH The Prince of Wales after his Graduation Parade, 20 August 1971 *l to r:* Air Vice-Marshal F D Hughes (Commandant): Air Chief Marshal Sir D F Spotswood (Chief of the Air Staff): HRH The Prince of Wales: HRH The Duke of Edinburgh: Air Marshal Sir L Mavor (AOC-in-C Training Command)

33 Roll of Honour in St Michael's Church

accommodation, educational facilities and amenities to enable it to fulfil its role of training permanent officers of the Royal Air Force.

The Working Party received additional guidance from AMSO that the provision of a permanent church was to be considered but the desirability of Technical and General Duties cadets being trained together was not to be considered by them. In this last caveat we see the emergence of one strand in the highly complicated web of discussions that were to surround Cranwell in the early 1960s. The Melvin Committee made a detailed study of the requirements of the College and the Air Council in due course accepted its proposals in principle for the construction of:

a. a new instructional block;

b. the fourth wing in the College;

c. a SNCOs Mess and quarters;

d. the modernization of airmen's barrack blocks;

e. a new Church of England church.

The projects for a gymnasium and swimming bath and for an Airmen's Club were postponed for the time being. The proposals for syllabus revision were not so readily accepted by the Air Council who asked for a series of additional studies to be made of the implications of such proposals as extending the course to four years and raising the academic standards required for entry to the RAF College. The Melvin Committee had indeed sounded a note which was to be heard frequently in high level discussions. The bone of contention was the Cranwell syllabus.

The Cranwell Syllabus

Post-war planning decided on a Cranwell flight cadet course of eight terms for about two and a half years total duration. This pattern was followed by Nos 45 to 68 Entries between 1947 and 1956. During this time Mr Patrick Johnson was the civilian Director of Studies and his tutorial staff consisted mainly of uniformed officers of the RAF Education, General Duties and Technical Branches. On 1 January 1953 Patrick Johnson was succeeded by Mr Antony Constant, MA.[11] One of his difficulties, an inherited one, was that at least every five years he had to deal with a complete changeover of his uniformed staff because of the personnel, posting and career requirements of the

Education Branch. In addition to the accommodation problems already mentioned, the Director of Studies suffered the disadvantage of being housed in an administrative building removed from the teaching blocks for science and the humanities which themselves covered a dispersed, straggling site.

The Hartley Report in 1952 had spelled out very clearly the short-term and long-term objectives of the Cranwell training. The short-term aim was to produce good junior officers and proficient pilots and navigators. The long-term aim was 'to provide the quality of education and officer training which will enable the Cranwell graduate to develop his powers and his facilities to meet the demands of progressively higher rank. He must be led to cultivate the capacity for logical and objective thinking, firm judgement and clear expression looked for in a good staff officer and a competent senior commander. He should be sufficiently an educated man to comprehend the broad as well as the narrow professional view.' To harmonize these two objectives was difficult enough. The task depended on two overriding factors: one was that the length of the training course was adequate; the other was the intellectual level of the cadets on entry and the academic level of the instruction imparted to them at Cranwell. There were signs that problems would arise over both these requirements. The Melvin Committee noted that the RAF officer of tomorrow would have to deal with scientists and specialists in every field as well as with senior civil servants many of whom were university trained. They were also of the opinion that one of the reasons why Cranwell was not attracting sufficient numbers of young men of the academic and personal qualities required was that headmasters considered the level of instruction at Cranwell too low for boys who were capable of reading to degree level.

From No 69 Entry onwards the cadet course was extended to three years. A further complication was added when the number of entries annually was reduced from three to two beginning with No 74 entry in January 1956. By the time of the inspection of the academic syllabus and the College by HM Inspectors of Education in May 1958 the Cranwell syllabus occupied three years of three terms per year. The first year was devoted mainly to academic work. Terms four to six were given over mainly to basic flying training. In terms seven to eight the cadet moved on to advanced flying which from the mid 1950s meant jet flying in the Vampire T11. The first navigator cadet had appeared at Cranwell in 1956. By 1959 arrangements had been made to give them jet familiarization by detaching them for a period to No 2 ANS at RAF Thorney Island. This was another disruptive factor in programme planning. The final term was spent almost entirely on flying training. This complex of activities included also ground training in Service subjects. The demands of night flying and compulsory rest periods and the necessity to recoup flying time lost during bad weather tended to make coherent academic studies difficult and frustrating. Cadets were tested on entry to the College in English, Science and Mathematics. They were then graded by ability into four groups for both Sciences and Humanities and were examined at the end of the second, fifth and eighth terms. They all took the same examinations. A tutorial system was used and the cadet was supervised in the preparation of a special subject in the last year or so of his course. It is not surprising that there were conflicts of interest as the staff attempted to get maximum effort for their own activity from the captive cadet.

The Air Council relied for some forty per cent of its General List permanent commissioned officers on the training at Cranwell. Suspensions from training for whatever reason were carefully examined and analysed. 715 cadets out of 3042 were suspended from training over the period 1947-73.[12] The reasons were as follows:

Medical	59
Academic	83
Character	89
Flying	231
Voluntary	253

Apart from the number of voluntary withdrawals from training, the relatively low percentage failure for academic reasons is noteworthy. Cadets obviously could conclude also that to devote less effort to academics was unlikely to have the disastrous results of neglecting flying or officer training.

From 1956 onwards other changes had taken place. Flight cadets no longer had to serve two terms with airman status in Junior Entry. They were given officer cadet status from the outset. The two entry per year system also introduced passing out parades in July and December. One less passing out parade meant the saving of productive time because the Cranwell graduation parade demanded hours of careful rehearsal. The pattern of a three-year course in which the first year was mainly academic studies, the second year given over to basic flying training on the Piston Provost and the third year to advanced flying on the Vampire lasted until 1962. It was however very much the subject of frequent discussion at Air Council level in the same way as that body also monitored the progress of the new instructional and domestic buildings authorised on West Camp. On 6 January 1959 AMSO speaking to a paper written jointly with DCAS outlined the Cranwell syllabus for the benefit of his fellow Council members as follows:

> Flying. Cadet pilots receive 45 hours navigation training in Valettas during their first three terms followed by 140 hours basic flying training on Piston Provosts in terms 4, 5 and 6 and 130 hours advanced flying training on Vampires in terms 7 and 8.
> Academic Studies. (Aeronautical Science, Engineering and the Humanities). The present syllabus covers 1008 instructional hours and the humanities syllabus 426 instructional hours. An unspecified part of the total private study time of 399 hours is also devoted to scientific and humanistic subjects.
> General Service Subjects. The present syllabus covers 740 hours.[13]

But the Air Council were thinking in broader terms about the Cranwell syllabus. They were concerned that the future permanent list officer of the GD Branch should be educated to a level beyond that of an 18-year-old entrant though the Chief of the Air Staff was alarmed at the idea of a four-year course at Cranwell as students. He was aware of the competition of the RN and the Army for good material from schools. The Air Council agreed in 1959 that the educational standards for entry to Cranwell should be raised.

The press release announcing these changes indicated that future entry qualification would be by a General Certificate of Education (or Senior Certificate in Northern Ireland) with passes in English Language, Mathematics, a Science subject or a foreign language and two other subjects. Two passes at A level in subjects other than English Language were required but Handicrafts, Domestic Subjects, Art and Music were excluded.

This meant that the Civil Service Commission examination was to be discontinued. The standards were to be for those candidates direct from civil life. In March 1960 CAS was considering also the idea of cadets moving on from Cranwell to pursue a degree course. This would compare with the various schemes being strongly canvassed by the Royal Navy and the Army.

The great obstacle to all these schemes, as it emerges in Air Staff thinking, was that the longer the flight cadet was kept away from flying the more expensive the training system would become and the less incentive and enthusiasm the cadet would show. Already the Air Staff comment on the three-year course with its element of passenger flying and navigational experience only in the first year was: 'The first year with no flying training is bound to be a disappointment'. There was some ground for concern about the standard of flying training achieved under the existing regime at the College. The Operations Record Book for the period frequently reports that flight cadets have been recalled in the breaks for special flying training instruction because they were below the required standard in their performance.[14] The flying at Cranwell was sometimes compared disparagingly with the more concentrated training of supplementary list entrants at flying training schools in the same Command. There were even wider considerations and pressures beginning to influence the thinking of the Air Council at the turn of the decade. These will be discussed later.

The third problem of the 1950s has already been foreshadowed, ie, the quantity and quality of the entrants coming from the schools for a career as a PC GL officer through Cranwell. There was no doubt that numerical targets were not being met. In the years 1948-56 the average entries to the GD Branch from Cranwell trained officers was 83 against a planned annual figure of 132. The RAF Scholarship scheme had been started to assist boys at school who wished to enter the GD or Technical Branch. Up to 65 vacancies for GD training and 20 in the Technical Branch were made available in 1956. In 1954 Lord Trenchard and the Chief of the Air Staff had expressed concern at the decreasing number of ex-apprentices entering RAF College Cranwell. The average annual entry of ex-aircraft apprentices was three despite the award of a prize cadetship for Cranwell or Henlow for each Halton entry. At the turn of the decade though many improvements were in train at the College, the recruiting position could only be described as unsatisfactory.

11 *The Winds of Change – The 1960s*

In a real sense the 1950s and 1960s cannot be called separate periods. They formed a continuum in the evolving discussions that were to shape the future of Cranwell. It was as early as 1954 that the Air Council became concerned about one aspect of the way ahead. The RAF Technical College at Henlow formed in the late 1940s was proving a successful attraction to young men from the schools anxious to obtain not only a Service career but the chance of a university degree. Technical cadet schemes were introduced in 1952. Under one scheme a student could follow at Henlow a Higher National Diploma course of three years and one term leading to an HND in Mechanical or Electrical Engineering. Under the other, University Cadet scheme, after a year at Henlow, up to 20 cadets were selected annually for engineering degree courses at Cambridge, Oxford, London, Bristol, Southampton, Birmingham, Manchester and Liverpool. They went as commissioned acting pilot officers and became pilot officers after two years at university. At the university cadets remained under the supervision of the RAF Technical College and became members of their University Air Squadron. After graduation they returned to Henlow for a two term orientation course designed to prepare them to take up their first appointments as junior technical officers. At Henlow, given the right medical standard, cadets were given facilities for *ab initio* training on the Chipmunk. In subsequent service they could apply for training to wings standard.

The Air Council became deeply concerned that the General Duties and Technical Branches were breeding two different kinds of RAF officer in two establishments both very expensive in staff and equipment. It was also realized that the introduction of guided missiles and ballistic rockets into the armoury of armed forces would postulate a high level of scientific and technical competence in the officers responsible for operating such methods of warfare. The Inspector-General (Air Marshal C N Guest) was therefore asked in 1954 to investigate the desirability and practicability of amalgamating the Royal Air Force College Cranwell and the Technical College Henlow.[1] Among other things he noted that it could be assumed that the key men in aerial warfare of the future would be those who possessed not only a wide knowledge of operating techniques but also the capacity to supply a scientific approach to the problem. It was towards the production of this type of officer that the Royal Air Force must now aim and plans must be formulated to ensure that the training of officers to meet this requirement was on the right lines. It was at the cadet stage that the process must be commenced. He recommended the merger of the two cadet colleges, provisionally in 1962.

The Air Council discussed this report in January 1955. In the discussion the statement was made that Cranwell was not yet preparing the GD cadet even for the beginning of the increasingly complex age ahead and that the academic entry standards at Cranwell were too low. The Council asked a committee under Air Marshal Guest to produce an assessment of the likely general military trends in the future and to suggest certain assumptions against which entry standards, training and career planning hypotheses could be constructed. The committee produced an interim report on 3 June 1955 and a significant conclusion reached was that manned aircraft and conventional weapons in addition to nuclear weapons would continue in use in varying degrees in all operational commands of the Royal Air Force.

Into the many discussions and plans for the development of Cranwell at the College, at HQ Flying Training Command and at the Air Ministry was now injected the concept of a merger between Cranwell and Henlow—a subject which was nevertheless forbidden to the Melvin Committee two years later as we have seen. The pot was set boiling indeed by the publication of the Sandys White Paper on Defence.[2] It is worth recalling some of the statements of that paper to appreciate the disturbing impact they made on the Air Council.

> In recent years military technology has been making dramatic strides. New and ever more formidable weapons have been succeeding one another at an increasing rate. It has been clear for some

> time that these scientific advances must fundamentally alter the whole basis of military planning.
>
> The means of delivering these weapons [ie, nuclear] is provided at present by medium bombers of the V-class whose performance in speed and altitude is comparable to that of any bomber aircraft now in service in any other country. It is the intention that these should be supplemented by ballistic rockets.
>
> Fighter aircraft will in due course be replaced by a ground-to-air guided missile system.
>
> The Government have decided not to go on with the development of a supersonic manned bomber, which could not be brought into service in much under ten years.
>
> The RAF are unlikely to have a requirement for fighter aircraft of types more advanced than the supersonic P1 and work on such projects will stop.
>
> The large reduction in the size of the forces will inevitably create some surplus of officers and NCOs. Those whose careers have to be prematurely terminated will be given fair compensation and will be helped in every way possible to find suitable employment in civilian life.

Such sentiments could hardly be interpreted by young men and their schoolmasters as a clarion call to a Service career along traditional lines in the Royal Air Force.[3]

The Air Council returned to close consideration of the Cranwell/Henlow amalgamation when they discussed in 1960 a paper written jointly by the Commandants of Cranwell and Henlow on the possibility of merging the two colleges. The Air Council not unusually were caught in another dilemma. The Henlow Cadet course now of 4 years and two terms duration was in process of being recognised by the National Council for Technological Awards as qualifying for the Diploma in Technology (Engineering) in Aeronautical Engineering. A letter of recognition was received at the Air Ministry in July 1960 with the reservation that the recognition would be reviewed in October 1965. The Air Council therefore instructed AMSO in May 1960 to set up a Working Party on the Amalgamation of Training of GD and Technical Cadets.

This Working Party met under the chairmanship of DGO (Air Vice-Marshal W P G Pretty) and submitted its Report in November 1960. The Chairman acknowledged to AMSO that the main problem of the

Working Party had been how to amalgamate General Duties and Technical Cadets' training without prejudicing the recognition of the Henlow Cadets' course for the Diploma in Technology which was the external qualification corresponding best with the sort of course needed for purely RAF purposes. The members of the Working Party were unanimous in recommending that the amalgamation of the training of Technical and General Duties cadets should be achieved by moving to Cranwell virtually all the training now done at Henlow. This was Scheme E, the fifth of the Schemes considered and provisionally costed for the amalgamation of cadet training. The argument was that the move would cost only slightly more than the amount estimated to be required for new building at Henlow if the RAF Technical College remained there. The RAF Technical College was responsible for a variety of training courses for technical officers. These too would have to be moved to Cranwell. The Report also considered that recognition of the technical cadets' course for the Diploma in Technology need not be affected by the move. The Air Council took a decision on 15 December 1960[4] that the Technical College should move from Henlow as soon as the necessary measures to receive it adequately at Cranwell were complete. Planning for the merger would, however, have to be superimposed on the plans and building developments already in train at Cranwell. For example, AMSO pointed out in July 1960 that the Air Council had approved the plans for building a new Assembly Hall, Swimming Pool and Gymnasium which would go ahead unaffected by the possible merger of the colleges.

The nature of the syllabus at Cranwell also remained a parochial matter and the subject of much controversy. In January 1959 following the Melvin Committee's recommendations the Air Council were being asked to agree that the Cranwell course should be of three years duration covering only basic flying training; that advanced flying should be undertaken in a fourth year at a separate school and that further study of the nature and purpose of the syllabus should be made. At Cranwell certain cadets were being encouraged to pursue their studies to degree standard. They were entered for external degree examinations of the University of London. There was undoubtedly much enthusiasm for the extra-mural visits, field work and lectures which broadened considerably the scope of work at Cranwell. The Operations Record Book for August, 1961 records that a camp to carry out geographical field work under the guidance of staff from the University of London was held at the Study Centre at Malham Tarn,

Yorkshire, for flight cadets reading Geography in preparation for the BA (General) Degree of the University of London. But the overall course length remained three years and the cadet had to be a very superior person indeed to learn to fly, to learn how to become a potential senior officer and to take a degree all in three crowded years. Some managed to do it but by the early 1960s the RAF was not recruiting that sort of top grade schoolboy, the upper 10 per cent of the school population. The majority were on the advice of their mentors seeking their career pattern through civilian universities.

Members of the Education Advisory Committee for the Royal Air Force visited Cranwell on 23-24 November 1959 to discuss a revision of the syllabus and in 1962 the old syllabus was changed to one of three streams: A Stream—General Studies, B Stream—Science, C Stream—Humanities.

This recognized the opinion of the Melvin Committee that if Cranwell were to fulfil its long term objective, cadets must be given the equivalent of a university education and this education must have science at its core. The course at the College from 1959 appeared indeed to be geared to the search for external academic qualifications in keeping with the current trends in national policy for higher education. Advanced flying was to be removed from the syllabus as a stage in training at Cranwell. The arrival of the Jet Provost in 1961 facilitated that decision and the Vampire T11 was phased out in April 1962.

The Air Council was being subjected meanwhile to more advice and pressure, often conflicting. AMP reporting on a recent meeting of the Principal Personnel Officers had reported to his colleagues the War Office's proposal to award Ministry of Defence Scholarships to suitable candidates who would join one of the Services after graduation from their sponsored studies at a university. The Graduate Entry Scheme was being born. Then came one of those startling 180° turns of thinking that appeared so bewildering to onlookers at the time. The Air Council studied a paper on the Cranwell syllabus[5] in which it was recommended that the College syllabus should be orientated towards full professional training for pilots and navigators including advanced flying. The paper also proposed yet another Working Party to make recommendations for the content of the Cranwell syllabus, including the level to which instruction should be taken in ground subjects. It further proposed that the revised syllabus should be introduced (together with any necessary change in entry requirements) at the time of the Cranwell-Henlow merger.

The Air Staff thinking was that in the long term the function of Cranwell might be to give professional and flying training to those who had already obtained degrees in civil life. Experience though brief had shown that it was impractical for Cranwell to attempt to offer academic training comparable to that of a university and at the same time provide satisfactory professional training. For some years to come there would be a need for officers with university degrees and also for those of lesser academic ability but with the required qualities of character and the ability to absorb professional training. The former category would increasingly be provided by the University Cadetship Scheme; it would be the main function of Cranwell to provide the latter. Whether it would be feasible to recruit the latter was left to speculation.

The Working Party on the syllabus was set up under ACAS (Training) Air Vice-Marshal P D Holder in November 1962 and submitted its Report in August 1963. On 29 August 1963 the Air Council decided that the aim of Cranwell was to provide the best possible all-round training to fit an officer for a full career in the Royal Air Force. A note of realism and alarm was sounded in the minutes since it was appreciated that only three courses had passed out from Cranwell since the 1959 arrangements were introduced and headmasters and others with influence in the education field might conclude that the RAF authorities did not know their own mind. The Holder Syllabus, as it became known, was to be introduced in October 1964 with the induction of No 91 Entry. This would mean a reversion to a course of about two and a half years duration. The effects of the publication of the Robbins Report in 1963 and the subsequent deliberations of the Melville Committee bearing on the question of recruitment of graduates to the Armed Forces will be considered later.

Discussions about the future of Cranwell did not progress smoothly and yet the College continued to function in a remarkably normal manner. The deep sense of tradition and dedication that kept Cranwell to its task, however confusingly that was defined, was greatly reinforced when in April 1960 HM The Queen graciously consented to become Commandant-in-Chief of the Royal Air Force College. On 25 July 1960 when she was the Reviewing Officer at the graduation of No 77 Entry, the Queen presented a new Colour to the RAF College to replace the one presented by her father in 1948.

Although the College was now to be overtaken by another change of syllabus philosophy and by the merger with Henlow, a number of long-planned building programmes were coming to completion. On 13 May 1960 the wall leading to the new fourth wing of the College was ceremonially breached and the building was made symmetrical and complete. Work was begun on the new Church of England church and the College Museum was opened for the visit of The Queen and HRH the Duke of Edinburgh. Basic Flying Wing was attached to RAF Spitalgate while major work was carried out at Barkston Heath so that Jet Provost flying could take place there as well as Cranwell. The contractors did not hand Barkston Heath back until February 1961. In addition to a new Instructional Block to the west of College Hall and a swimming pool and gymnasium, a new church and a modernized NAAFI shop were being constructed. Other major projects in the offing were a new Airmen's Club and a new Sergeants' Mess. On top of all this the necessary building plans in connection with the move of Henlow to Cranwell were being discussed. Members of the Working Party under DGO (Air Vice-Marshal W P G Pretty) visited Cranwell with the architect Mr Mitchell and agreed with the Commandant, Air Commodore D F Spotswood, that the best site for the instructional block for engineer training, provisionally to be called the Institute of Technology, should be in East Camp adjacent to the building known as the Taj Mahal. They also decided that the new Student Officers' Mess should be built on the orchard site also in East Camp. The polarization of West Camp and East Camp was thus confirmed in May 1961.

Work progressed rapidly. The Roman Catholic Church of St Peter was modernized in this period. The new church of St Michael and All Angels was dedicated on 1 June 1962 and the Memorial Chapel was transferred from the College to the new Church, where its original style and beauty have been faithfully preserved. The King's Colour presented in 1948 and the Queen's Colour presented in 1960, both well-worn, hang above cabinets containing Books of Remembrance placed between the arches separating the Chapel and the Nave of the Church. The Lectern, on which are inscribed the names of men who died at Cranwell during the First World War, was brought from the old Hangar Church and is among many gifts reflecting so much of Cranwell's past and richly embellishing St Michael's today.

On 4 October 1962 Sir Frank Whittle, a former aircraft apprentice and flight cadet at Cranwell, opened the new Instructional Building which became known as Whittle Hall. Its main feature, perhaps, was a magnificent assembly hall with good audio-visual aids where a variety of activities could take place including the award of wings and prizes. Work on the so-called RAF Institute of Technology by Messrs A Monks and Co Ltd was planned to begin in September 1963 and to be completed in September 1965. To add to the growing community facilities a new Cranwell County Primary School was built and opened on 14 May 1963 on the North airfield. It was built by Messrs Bosworth and Wakeford and authority was given in May 1961 for an additional 152 Officers' Married Quarters to be built at Cranwell.

From No 81 Entry onwards flying training at the College was to be carried out on Jet Provost aircraft. It was to cope with a somewhat bewildering situation on the academic side that a new Director of Studies was appointed in April 1960. Mr J A Boyes who had been Headmaster of Kendal Grammar School from 1955 to 1960 took over from Mr Antony Constant. The new Director could indeed be forgiven if he believed that he had inherited a mare's nest.

Plans for the amalgamation of Henlow and Cranwell became overriding factors. In May 1961 a committee had been appointed under Air Marshal J D Baker-Carr to examine the future organization and structure of the RAF College when Henlow had moved up to Cranwell. Their interim report was submitted to the Air Council by AMSO in February 1962. One of the firm assumptions given to the committee was the maintenance of the Diploma of Technology standard for the Henlow Technical Cadets course.

What had been proposed by Director-General of Organization's Working Party in 1960 was that virtually all the training now done at Henlow should be moved to Cranwell. There were many dissentient voices—at Cranwell, at Henlow and within the Air Staff. A typical opinion was that Cranwell had always dealt with cadets only. It compared and competed athletically with Sandhurst and Dartmouth. To amalgamate the RAF Technical College and the RAF College could destroy institutions as such. Financial savings and staff economies it was said, appeared to have dominated the reasoning behind the proposed merger. Doubt was also cast on the accuracy of the arguments which had found for Scheme E as opposed for example to one or more of the other schemes proposed as a basis for the merger.[6] The idea of cadets training in close community with student officers at Cranwell also aroused serious concern. Memories of the unpopular graduate entry scheme of 1934-36 may have influenced thinking. Such an environment would undermine the disciplinary atmosphere in which cadet training had always been conducted. In other words a university campus and a military academy environment were not compatible in some senior officers' opinion.

But the decisions had been taken. These decisions moreover, soon appeared to be endorsed by two important reports, one the Robbins Report of 1963 and arising from it the Report of the Melville Committee.

The Robbins Report dealt with the dramatic expansion of opportunities for university education available to school leavers with the minimum qualifications of A level passes—such as were demanded for entry to Cranwell. It was estimated that the proportion of each yearly age group obtaining the minimum educational qualifications for university entrance would rise from seven per cent in 1961 to thirteen per cent in 1980. The increased opportunities and the expectations of school leavers reinforced by urgings from their schoolmasters and parents would lead young people to examine closely what the Services had to offer them in the way of tertiary education. It was felt generally that the Services should either recruit young men and women with a recognized academic qualification or—if they were candidates for a General List or permanent regular commission—require them to complete a degree course after entry. The Defence Council in July 1964 decided to set up a very powerful committee chaired by Sir Ronald Melville Second Permanent Under-Secretary at the Ministry of Defence with the following terms of reference:

> To examine the requirement of the Services for trained manpower over the next 15 to 20 years and the associated implications for the training of officers and men in the light of national developments in education and in the interests of increased economy and efficiency. The Committee should take account of the studies being carried out separately on the scope for inter-Service rationalisation of training.

One must note here the watchwords of economy and inter-Service rationalization, twin altars of gods to whom many sacrifices have been made in recent years.

The Air Force Department set up its own Working Party under DGT which prepared a paper for the Air Force Board to consider in June 1964 before evidence was submitted to the Melville Committee. The discussions were thoughtful and thorough. Broadly the paper acknowledged that the RAF must seek to raise the entry standards for all branches of officers and also for the more highly skilled airman trades.[7] The aim should be eventually to make the possession of a degree or its professional equivalent the basic requirement for direct recruitment to the General List in all branches. However, to maintain a balanced officer corps within each branch, General List commissions should continue to be awarded to officers on the Supplementary List who would themselves have been commissioned either direct from civil life or from airman rank. Directly recruited General List officers in the General Duties, Technical, Secretarial and Equipment Branches and in the RAF Regiment[8] could obtain their degrees in one of two ways—either at a civilian university or at a degree-awarding Cranwell. There was a case for advocating Cranwell's evolution into a degree-awarding body. It was suggested that this subject should be reviewed in five years' time. For the same reason it was suggested that the idea of a Service university for the Royal Navy, Army and the RAF should also be considered in five years' time. In that connection Cranwell could be thought of as an Air College in a defence university. In the meantime it was proposed that the number of University Cadetships[9] be increased to 40 per annum and more if the scheme seemed justified. It was also suggested that an approach be made to the Council for National Academic Awards to attempt to secure some recognition in degree terms of the content of the current GD flight cadets' course at Cranwell.

The Air Force Board Standing Committee in its Conclusions 5 (64) on 24 June 1964 made up its mind that the all-graduate approach with Cranwell reverting to its original role of providing the necessary professional training was the cheapest financially and most promising solution for the Royal Air Force. If a choice had to be made between Cranwell becoming part of a defence university or becoming affiliated to a civil university, the latter course seemed preferable at this stage, certainly on grounds of economy.

The Melville Committee made its detailed and thorough Report to the Defence Council in April 1965. Some extracts will assist the reader to gauge its impact on Air Staff thinking. It acknowledged that the RAF had university cadet schemes in operation plus the few degree students at Cranwell who managed to read for an external degree. It stated that to send students to read for a degree at a civil university after being trained at Cranwell would largely waste their previous flying training. It also noted that it cost the Services much less to pay for an officer at a university than to provide academic training to degree level in a Service college. It advocated that the Services should take as many officers from the universities as they could get. For the forseeable future they were unlikely to be able to recruit more than they could absorb. It recommended further that cadetships in particular specializations for RN and RAF officers should be removed thus widening the scope of the cadet scheme. If this were done, on fairly optimistic estimates, the RAF might be able to obtain from the universities by about 1980 their full requirement of officers for GL permanent commissions from civil life, and so replace the cadet entry through Cranwell altogether. (This would mean recruiting about 105 to 115 in the GD Branch of whom 75 or so came from Cranwell each year and 74 for the Engineering Branch, of whom 20 should be graduates. About 40 per year came from Henlow courses).

The Report also touched on the concept of concentrating all Service technical degree training at Shrivenham and the necessity of thinking what could then be done with the £2,250,000 already expended on new facilities at Cranwell. It advocated finally that the pattern of professional training at Service colleges in the UK should not be distorted for the sake of obtaining a degree qualification in general subjects. It did however recommend that the RAF pilots and navigators course might possibly qualify for a Pass degree recognized by CNAA without an extension of the course length of three years.

One can add here a comment that the UK Government short of resources and cutting its defence costs was never in a position to take the kind of imaginative, national leap forward that has made the United States Air Force Academy at Colorado Springs (founded in 1954) one of the major degree-awarding academic institutions in the USA.

The British attempted solution was to set on foot an enquiry under Professor M E Howard (at that date Professor of War Studies, King's College, London) and Mr C R English, Senior Chief Inspector at the Department of Education and Science. In November 1965, the Secretary of State for Defence (Mr Denis Healey) gave them the following terms of reference:

a. To recommend what measures might lead to greater efficiency and economy in Service training at undergraduate and graduate level.

b. To examine whether it is desirable and practicable as a long term measure to concentrate in one place the whole or part of the training resources of the Service Colleges at Dartmouth, Manadon, Greenwich, Sandhurst, Shrivenham and Cranwell.

c. The examination should take account of the Report of the Committee on the Implications for the Services of National Developments in Education. It should also take account of the implications in terms of capital expenditure of changes in present arrangements.

The last sentence brings us appropriately back to our home base. What indeed was happening at Cranwell in November 1965 when the Howard-English team received their briefing from the Ministry of Defence?

12 *The Cranwell-Henlow Merger*

It would be unjust to those officers and cadets who served at Cranwell in the late 1950s and early 1960s to present their story as one of continual turmoil and planning changes. Cranwell, if one reads its Operations Record Book, appeared on the surface to be astonishingly normal. On 11 January 1962 HQ FTC agreed to change the title of the Unit Commander to that of Station Commander though on all official correspondence he was addressed as Officer Commanding RAF College Unit. The pattern of organization was now firm and stable. This was the period of visitations to Cranwell. A continual stream of representatives from foreign Service academies, industrialists and lecturers from universities and polytechnics came to see the developing facilities of the College. On 26 April 1963 the Ministry of Defence staged a conference of headmasters at Cranwell. A total of 158 came from schools and colleges and met 25 senior officers of each of the three Services. The Secretary of State for Air, the Rt Hon Hugh Fraser, opened the conference—a sign of the seriousness with which the authorities viewed the general recruiting situation.

The cadets themselves as always were engaged in reciprocal visits with Salon, Colorado Springs, Dartmouth and Sandhurst and as always managed in their studies to secure detachments to fascinating and instructive operational units of the three British Services. This was the most stimulating period also for higher educational studies and cadets were entered for the external degree examinations of London University, for the Civil Service Commission Interpreter and Linguist examinations mostly in French and for Parts I and II of the AFRAeS examinations. Cadets also took part with success in such events as the Ten Tors Walk on Dartmoor, the Nijmegen Marches in Holland and the regular leadership and initiative exercises, tough and demanding physical effort, in Yorkshire. Overseas camps included those in the Troodos Mountains of Cyprus and 'King Rock' exercises in Germany. They also began to participate in the Trenchard Award Scheme.[1] One such adventure planned was 'the under-water survey of the ancient port of Sabratha near Tripoli.' Nor was invention lacking nearer home. The RAF College Hovercraft CHI was the centre of interest in August 1964 at the Sleaford Agricultural Show where an oil leak was temporarily but effectively stopped by the judicious placement of chewing gum. During this time visitors to the College had included parties of candidates from the RAF Selection Board. These visits were usually on Sunday afternoons. The Selection Board which occupied Daedalus House from July 1962 to April 1964[2] moved down to RAF Biggin Hill and the Equipment and Secretarial Wing took over some of its other accommodation.

The local contacts also continued to be fostered, at Remembrance Day ceremonies, graduation parades and events such as the dedication of the new Church of St Michael and All Angels by the Lord Bishop of Lincoln in June 1962. Cranwell may have had its internal difficulties but to the visitor it appeared to be very much the impressive unit it had always been.

However, the shadow of future changes clouded most of the thinking of the authorities at Cranwell. In March 1962 the first sketch plan of the so-called RAF Institute of Technology was produced for comment. In January 1964 an important conference took place at Cranwell on the merger with Henlow. In November 1965 the Inspector-General and his staff officers visited Cranwell to inspect and discuss the progress of the merger. What were the plans for this great enterprise? The Minister for the RAF had agreed that the move should be undertaken by the Autumn Term of 1965 when the first technical cadets would report direct to Cranwell. Those already at Henlow would move to Cranwell in January 1966.

The organization at Cranwell to receive the combined colleges was the subject of much discussion in the Air Force Board[3] in the 1 (64) and 3 (64) Meetings especially. Agreement was soon reached that the post of Commandant should be of air vice-marshal rank. Eventually decisions were taken that there should be an Assistant Commandant (Cadets) who would be responsible for some 500 cadets in all. An Assistant Commandant (Technical) would be responsible for the RAF

Institute of Technology covering post-graduate training for some 350 officers but also providing instruction for cadets in certain technical subjects. Responsibility for flying training would rest with the Chief Flying Instructor under the Assistant Commandant (Cadets) but the Station Commander would be responsible for aircraft servicing and normal airfield services. A Senior Director of Studies of high quality and appropriate experience would be appointed from civilian life with the equivalent status of air commodore—the same as the Assistant Commandants. There would be two Assistant Directors of Study, one for Whittle Hall and one for the RAF Institute of Technology. A new appointment, Officer Commanding Cadet Training, would be established in the rank of wing commander and he would supervise the activities of the four squadron commanders in College Hall. The name Institute of Technology did not find favour within the Air Force Board and in June 1964 the Minister (RAF) agreed that the name should be changed to Trenchard Hall. The foundation stone of the building was laid by the Chief of the Air Staff, Air Chief Marshal Sir Charles Elworthy, on 26 June 1964 and the opening ceremony was performed on 17 May 1966 by The Rt Hon Viscount Trenchard of Wolfeton, the only surviving son of the late 1st Viscount Trenchard, Chief of the Air Staff 1919-29. The contractors indeed were still in the building when on 1 January 1966, 104 officer students from RAF Technical College Henlow began instruction at Cranwell. They were joined by 36 officer students for newly assembled courses. A Merger Celebration Dinner was held in College Hall on 3 February 1966, attended by the Minister (RAF), Lord Shackleton, and by RAF commanders-in-chief from the commands. A Press Day previously had publicized the merger.

The situation at Cranwell now was that the Commandant and his personal staff were in the Headquarters Building where the Station Commander and his staff also had their offices. The Director of Studies too was located there. In College Hall were the Assistant Commandant (Cadets), the Wing Commander (Cadets) and the senior entries of A, B, C and D Squadrons. Basic Studies Wing under the Assistant Director of Studies (Sciences and Humanities) was in Whittle Hall and in Trenchard Hall were the bulk of the technical training facilities with an Assistant Director of Studies (Technical) under the Assistant Commandant (Technical). The fact that cadets had to attend certain classes in Whittle Hall and others in Trenchard Hall as did student officers on engineering courses meant a highly complicated system of transport had to be devised to make the programme feasible.[4]

There were now four messes—the Station Officers' Mess, College Hall Mess, the new Junior Officers' Mess in East Camp and the Junior Cadets' Mess in West Camp. Cranwell had become a very complex affair. Nomenclature to say the least was confusing. The naming of Whittle Hall where the flight cadets spent most of their time on academic work had pre-empted the naming of Trenchard Hall. A case could be made for switching them to achieve more fidelity to historical relevance. The mixing of cadets and student officers which had been viewed with such foreboding by many was now almost complete. To run the Station as a unity was the firm resolve of the Commandant, Air Vice-Marshal IDN Lawson appointed on 28 December 1964 and of the Assistant Commandant, Air Commodore N C Cameron (1965-66). But the factors of geographical dispersion on the camp and the totally diverse nature of the principal occupations of aircrew and engineers made the process of unification extremely intractable. The situation could hardly have been assisted by the appointment of a new Director of Studies, Dr George Tolley, who after replacing Mr J A Boyes in August 1965 departed from Cranwell in early 1966 to a presumably more congenial appointment.

The year 1966 was a difficult one for the College. The flight cadets were working to different syllabuses according to their entry number. The Holder Syllabus had been introduced for No 91 Entry in October 1964. This syllabus had reduced the Cranwell course to just under two and a half years instead of the previous three-year course. The year was now to be divided into two terms of 20 weeks each. In Term 1 the syllabus comprised Officer Training, Academics and Chipmunk flying for pilot cadets. Non-pilots were to have additional academics during the Chipmunk phase. In Terms 2 and 3 the syllabus consisted mainly of academic studies with a continuing thread of officer training. In Terms 4 and 5, pilots were flying the Jet Provost and doing officer training on a continuous basis. Non-pilots were engaged on specialist training and on officer training on a termly basis.[5]

The work of the Programme and Plans secretariat was indeed complicated and a source of endless improvisation and adjustment. This statement is heartfelt by the author who on 24 October 1966 was appointed as the Assistant Director of Studies in Whittle Hall. While ostensibly attached to the Department of Cadets and willingly identifying with their activites, he also had to oversee the relevant academic studies in Whittle Hall of some of the officer students from Trenchard Hall. In 1966 they came from the following melange of courses:

No 33 Technical Officer (Graduate) Electrical
No 33 Technical Officer (Graduate) Mechanical
No 5 Technical Officer (Standard) Electrical
No 5 Technical Officer (Standard) Mechanical
No 12 Technical Officer (ex Cadet Entry)
No 10 Advanced Weapons Course
No 34 Technical Officer (Graduate) Preparatory Course
No 3 Technical Officer (Supplementary List) Electrical
No 3 Technical Officer (Supplementary List) Mechanical
No 22 Short Data Handling Course
No 14 Engineer Officer (Branch) Orientation Course
Non-MOD Maintenance Engineering courses for foreign Services

It was indeed a bewildering campus. On 20-21 September 1966 Council members from the Council for National Academic Awards visited Cranwell to inspect the facilities. They were not happy with what they observed. In effect they gave the authorities a year's grace to make certain amendments to the syllabus and tutorial programme to qualify for the continued academic recognition of the former Henlow courses. Other visitors, harbingers of complicated proposals, were Professor Michael Howard and Mr Cyril English in March 1966. They came as part of their inspections of Service colleges preparatory to advising the Minister of Defence on the feasibility of rationalizing and bringing together Service academies of higher education. In the meantime the Air Force Board had decided[6] that advanced flying training would not now take place at Cranwell as envisaged under the Holder Syllabus but would be given post-Cranwell at the appropriate Advanced Flying School or Advanced Navigation School.

A previous delegation to Cranwell led by Rear-Admiral C R Darlington in October 1964 had been concerned with the Implications for the Services of the National Developments in Education. One hopeful proposal in the Melville Report which had engaged the attention of the Air Force Board was that the Council of the CNAA might recognise the academic course taken by aircrew cadets at Cranwell for degree purposes. It soon became clear that the CNAA would expect a course to contain 1000 academic hours, including private study as well as tuition, each year for three years. 2500 to 3000 academic hours would constitute a full degree course. Since the Holder Syllabus envisaged only 1940 hours in two and a half years it was clear that CNAA recognition was unlikely. Their decision to this effect was conveyed to the Air Force Board on 17 March 1966[7] by the Director of Flying Training (DFT). Hence the decision taken to modify the Holder Syllabus putting advanced flying to AFS and ANS respectively rather than undertaking an academic course at Cranwell of two and a half years followed by the equivalent of a basic flying training course on Jet Provosts.

The Commandant in the years 1966-68 was therefore faced with the following problems over and above his main concern to make his unit a single cohesive command. He had to face a series of graduation ceremonies as various entries of cadets' of the General Duties, Engineer and other ground branches came to the end of their courses. He had to ensure that the engineering courses for technical cadets were so reorganized that they would secure the approval of the Council for National Academic Awards. He had to supply evidence and take part in discussions about the Howard-English proposals which would have fundamentally altered the nature of Cranwell. He had also to heed the changes of thinking which in the Air Force Department and at HQ Flying Training Command were taking place following the findings of the Melville Committee. It was fortunate that at this critical period Air Vice-Marshal T N Stack was appointed as Commandant. He was steeped in Cranwell tradition and determined to ensure that no change occurred unless higher policy demanded it. The College needed to be kept on an even keel if the effects of the Cranwell-Henlow merger were to be absorbed.

The advent of the technical cadets and officers to the magnificent new buildings in Trenchard Hall brought an even greater volume of visitors. These included eminent industrialists, academics and representatives of foreign Services. A new relationship was formed with the German Air Force's technical academy at Neubiberg, near Munich.

The Operations Record Book for the period shows the kaleidoscopic activities characteristic of a lively station. In January 1967 two sharply contrasting and unusual events were recorded. A disastrous fire burned down the Workshop Technology Building (Block No 260). Seven fire brigade appliances were needed to help the Unit Fire Services to fight the blaze from 1728 hrs to 2330 hrs on 17 January. In the same month the College was delighted to learn that Flight Cadet P B Glover had been picked to play for England against Australia at Twickenham.

Cranwell was a showpiece for visitors from far and near. These included 4000 spectators in June 1967 who came to see the Lords

Taverners Cricket Team and a Red Arrows' display and contributed £600 for the Duke of Edinburgh's Playing Fields Fund. The graduation parades also continued to draw distinguished Reviewing Officers and guests. A link with tradition was made when Sir Frank Whittle presented academic awards on 1 November 1967 in the hall that bears his name to Engineer officers graduating from Nos 9, 10 and 11 Entries.

A regular feature of the graduation parades was the flypast of aircraft to coincide with the arrival on parade of the Reviewing Officer. Prominent in such aerial displays at Cranwell and elsewhere were the recently re-formed Jet Provost Poachers[8] who in autumn 1968 were asked to fly a '50' formation at the SBAC Show at Farnborough to celebrate the 50th Anniversary of the formation of the Royal Air Force.

Local ties continued to be strong. By the courtesy of Captain Jeremy Elwes of Elsham Hall and the Lincolnshire Association an exhibition of Russell Flint paintings was staged in College Hall. A more sombre liaison took place when the Royal Air Force provided 500 gallons of fuel oil to neighbourhood farms for the sorry and gruesome task of burning stock slaughtered in the outbreak of foot-and-mouth disease of early 1968. The proximity of farming and flying was never more emphasised than when the airfield was put out of bounds and the wheels of aircraft and MT vehicles were sprayed against spreading infection. The station had its own worries since the RAF Equitation Centre had been opened at Cranwell in December 1967.

In June 1968 the RAF College was given Group status within Training Command[9] when that formation was constituted. The departure of Dr G Tolley had brought Group Captain (Acting Air Commodore) C E P Suttle, an officer of the RAF Education Branch with much experience of Henlow to become Senior Director of Studies at Cranwell. Additional to his worries about the CNAA degree was the limitation of his teaching staff by an Establishment Review in early 1967. It was apparent that the level of educational studies other than those of engineering cadets was to be very much lower than that already being achieved with entries prior to No 91. The same Review agreed to the establishment of an Engineering Management Wing in Trenchard Hall. The character of the flight cadet course began to alter perceptibly. The College was already feeling the impact of Whitehall thinking about the Howard-English proposals and those of the Melville Committee.

The Howard-English Report was presented to the Secretary of State for Defence on 29 July 1966. In Section 2 of the Report the authors affirmed:

> It can hardly be desirable that in two decades time, when some 15 per cent of the population of this country will be receiving courses of tertiary education leading to some form of academic qualification, the Services should be largely officered by men who received no formal education after the age of 18.

Briefly what the Howard-English Report proposed for the non-technical officers at Cranwell was that they should spend a first year at Cranwell on military and general service training with some elementary flying training in Chipmunks and some time on a unit for familiarization with the role of the Royal Air Force. Their second year would be an academic year not at Cranwell but at a Royal Defence College either at Greenwich or at Shrivenham in company with Royal Navy and Army cadets. Then would follow basic flying training at Cranwell and later advanced flying.

It was also proposed that a Royal Defence Academy should be formed and that the Engineering Department at Cranwell should come under that Academy. The initial flight cadet training and all flying training would continue to be the direct responsibility of Flying Training (later Training) Command.

This brief sketch will indicate that the proposals to bring together in an academic environment not at Cranwell cadets other than engineers would have spelled the end of Cranwell as it had been planned. Further, such were the Service training plans of the Royal Navy and the Army, the RAF cadets at the Royal Defence College would have been a year or two years younger than their other Service counterparts and with little knowledge of their Service in operations and with no deep experience of flying. It was not surprising that the Howard-English proposal won little favour in the Air Force Board. Although many position papers and conferences were prepared to discuss the content of an academic course at the Royal Defence College, the feeling of the Air Staff was that the Howard-English proposals were attempting a marriage of incompatibles. Each Service had a distinctly different role in its lower echelons and a different training requirement—particularly in officer training. Yet the Air Force Board could not evade the strictures of the Howard-English Report: 'Our judgement in this may be at fault, but we feel it would be quite improper to continue to attract young men into the armed forces with promises of educational opportunities which cannot be fulfilled.'

Having set its face therefore against preparing cadets for degree studies at Cranwell and having taken the decision to maintain the flight cadet course length at about two and a half years, the Air Force Board, in order to meet the demands of public opinion and the pressure from the schools and educational bodies for affording opportunities for degree studies for bright entrants, had to look for other ways of increasing the proportion of graduates coming into the Royal Air Force.

The advice of the Melville Committee on this was unambiguous. The faint possibility that the CNAA would recognize the existing GD cadets course for degree purposes had been wiped out. The advice, already referred to above, was that the RAF might be able to obtain from the universities their full requirement of officers for General List permanent commissions from civil life, and so replace the cadet entry through Cranwell altogether.

It was in the summer of 1968 that the staff at Cranwell received the news that the cadet entry system as it had been known was to be phased out. A new training pattern was to begin with No 99 Entry in September 1968. The Air Member for Personnel directed that the first year of training of that entry was to equip the maximum number of flight cadets to find places in civilian universities and colleges and on the engineering degree course at Cranwell. The process of run down had begun. The old cadet entries would peter out and a new Graduate Entry Scheme would begin. The achievement of recognition by the CNAA of the engineering degree course at Cranwell was a hollow triumph after all the strenuous staff work of the preceding year. The Engineering Standard Course was also to be phased out and with the demise of the CNAA degree course there would eventually remain only a short Maintenance Engineering Course plus an Initial Engineering Course for new entrants to the Engineer Branch.[10] The Advanced Weapons Course and later the Aero Systems Engineering Course would still demand teaching and technical training to a very high level.

Two eventual results became apparent. One was that the magnificently provided facilities in Trenchard Hall would be under utilized. The other was that a considerable number of very highly qualified Senior Specialist Teachers[11] of Mathematics, Physics and Engineering Sciences in the RAF Education Branch would no longer find work worthy of their intellectual capabilities when Henlow and Cranwell no longer existed in their former roles. The majority of these officers saw out the run down of the cadet entries at Cranwell and then left the Service for more congenial and lucrative appointments in civilian life. The other main consequence of the ending of the traditional cadet entries was the beginning of a divisive debate about the wisdom of the new policy for Cranwell. This debate continues and it is no part of this historian's task to do anything but reflect the situation as it rests. Judgement is for a later historian. In the meantime Cranwell continues as a vital element in a fighting Service.

It is not possible either, even ten years later, to attempt a balanced evaluation of the success of the Cranwell-Henlow merger. That union was so caught up in cross-currents of national educational developments and rationalization plans for the future of the academies of the three Services that the necessary period of calm and stability to allow the merger to take effect was never achieved. Before the effects of the amalgamation had time to be absorbed, Cranwell was launched on a radical new experiment which changed the character of life there. The Graduate Entry Scheme was at the heart of the new pattern of training at Cranwell.

13 *The RAF College 1970-1980*

It was tempting to conclude a history of Cranwell with the demise of the old cadet entry scheme: to leave the latest decade to the pen of a future historian. It would have made a rounded story but it would have had an ending like that of T S Eliot's Hollow Men—'Not with a bang but a whimper'.

The decade since 1970 has seen great changes at Cranwell. The present organization is foreign to those who knew Cranwell in the 1930s or in the two immediately post-World War Two decades. But elements of historical continuity are discernible and the location of Cranwell, its buildings and amenities continue as a moulding influence on new patterns of training.

It was the Air Member for Personnel, Air Marshal Sir Andrew Humphrey, who had the task of expounding to the Air Force Board his ideas on The Future of Cranwell. Arguing from the premise that the long-term aim, as agreed by the Air Force Board in 1964, should be to make the possession of a degree or its equivalent the basic qualification for direct entry to the General List of the main executive branches of the Royal Air Force, be planned for the gradual phasing out of the cadet entry and also for discontinuing the Cranwell engineering degree course. Graduates would be obtained from civilian universities and colleges, encouraged by suitable financial incentives and conditions of service. Planning projections anticipated recruiting 170/180 graduates initially, working up to 250 per annum in due course to allow for training wastage.[1] The four main executive branches were the General Duties, Engineering, Supply and Secretarial branches. The argument was also concerned with the use to be made of the new, expanded facilities and accommodation at Cranwell. Two main lines of development were suggested for consideration. Cranwell could become an initial training centre where almost every officer of the main branches of the Royal Air Force would spend a period of his training—to absorb the atmosphere of Cranwell. Or Cranwell could become a centre for advanced training particularly in aviation. The use of flying facilities at Cranwell presented particular problems of 'peaking' of student population with numerically small entries at fixed intervals in the academic year. The future use of Cranwell became involved also with the long-term economy programme of station closures. Planning indeed entered a period of pragmatism and expediency where a simple clear-cut philosophy of training however desirable could be maintained only by uneconomic use of resources.

One issue was resolved by the time that AMP presented AFB (68) 28 to the Air Force Board. At the invitation of Mr Denis Healey, Sir John Wolfenden, then Vice-Chancellor of Reading University, had considered carefully the total concept of the Royal Defence College. His views which are given in the Supplementary Statement on Defence Policy 1968 (Cmnd 3701)[2] gave the quietus to the total concept. The Air Force Board adopted Sir John Wolfenden's suggestion that the Services should use 'the existing arrangements for sending individuals to existing Universities for normal civilian degree courses.' This policy was in line with the recommendations of the Melville Committee and soon the idea of a Foundation Year course for officers of all three Services at Greenwich was also quietly dropped. It had always been less than popular with the Air Force Board.

The Chief of the Air Staff was very aware of the fundamental changes about to be introduced into the pattern of recruiting permanent commissioned officers and into the organization of Cranwell. He went to great pains to write personally to all senior RAF commanders to explain what was happening and to set before them the possibilities for the future. The options were that Cranwell could become an alma mater for all RAF officers or it could become the seat of advanced professional study in the RAF such as was currently conducted at the College of Air Warfare at RAF Manby. Either way, and however the debate was resolved, Cranwell would continue to be the Royal Air Force College in fact as well as in name.

There would be an uneasy transition period at the College. The Spring 1969 non-Engineer cadet entry was cancelled. There were to be no more Spring entries. Cadet entries to Cranwell would cease in the

Autumn of 1970. In fact No 101 Entry arrived for training as the final flight cadet entry on 28 September 1970. It was in the same month that No 1 Graduate Entry reported to the College. Prior to their arrival the first of the University Cadets Introductory Courses had taken place at Cranwell. They were to be held in future years to cater for the Easter, Summer (Main) and Christmas university cadet competitions. University cadets were to attend the couse for a week (later it became 14 days) to complete initial RAF documentation, be issued with uniform and other service clothing and be given some General Service Training (GST) and an introduction to the Royal Air Force and to the work of the University Air Squadrons. The latter was important. Treasury approval to the continued existence of piston-engined flying at various University Air Squadrons was integral to the new scheme of Graduate Entry recruitment. Indeed as the extracts given in Annex XIII will show, the constant monitoring of academic, officer and flying training progress was a principal feature of the university cadet entrance scheme.

The run down of the old cadet entries had already caused amalgamation of A and B Squadrons in College Hall into No 3 Squadron, with A and B Flights. C and D Squadrons merged into No 4 Squadron, with C and D flights. Eventually one squadron would be formed. The Graduate Entry Squadrons would be known as Nos 1 and 2 Squadrons. The priorities were changing. From 15 June 1970 the Department of Cadets was renamed the Department of Training with an Assistant Commandant (Training).

It was to a College in the throes of transition that HM The Queen and HRH Prince Philip came on 12 June 1970 to mark the Golden Jubilee of the RAF College. They had an especial interest in the welfare of the establishment since it had been decided that HRH The Prince of Wales was to begin his flying training on the Jet Provost at Cranwell early in 1971. No 1 Graduate Entry was to get off to a splendid start honoured by the presence of a Royal pupil, the Heir to the Throne.

Among the changes of scene to be recorded was the disappearance of the Varsity aircraft on 23 September 1970. Chipmunk flying had almost ceased in the previous year. At the turn of the year 1970-71, the JP5 arrived and a special flight of three aircraft, Golden Eagle Flight, was formed in readiness for the arrival of Prince Charles.

The administrative and security arrangements for the training of the Prince of Wales were meticulously planned. But before examining them in detail, account must be taken of other decisions which affected

RAF planning. In July 1969 the Air Force Board took note of the Report of the Hodgkinson Committee[3] which had been examining the structure of officers' careers in the Royal Air Force. This Report advocated the replacement of the former General and Supplementary Lists by a Single List for each officer branch of the Royal Air Force. It also recommended that career opportunities for the new graduate entrants should be so safeguarded that at least 80 per cent of them would secure selection to the 'command stream', ie, to pursue careers to age 55 beyond the 38 years of age or 16 years of service break point. The Board at this time also agreed that the RAF should be reduced to no more than four main functional branches and about six professional branches.

The Single List concept countered any suggestions of elitism, of dividing the sheep from the goats in officer career planning by introducing a strong element of democratic career competition. The Air Force Board saw fit when it next considered The Future of Cranwell in AFB (70) 9 to reaffirm the aim of Cranwell unchanged as: 'To train the future leadership of the Royal Air Force.' But while firmly putting into abeyance a decision to move the Officer Cadet Training Unit from Henlow to Cranwell, the Board sounded a warning note that the College could well become a loose grouping of separate schools, instead of the single college with several faculties which would be the ideal.

It was about this time that it was realised that the Single List implied that ideally there was only one kind of officer—the Royal Air Force officer.

Hence the thought arose too that graduates and non-graduates should undergo the same Initial Officer Training, following the same syllabus. This idea was bruited at an Air Force Board meeting in 1971 and it was also stated that the period of exposure to the Cranwell influence for Graduate Entrants should be as long as possible. It was becoming apparent also that some mixing of non-graduates with graduates might have to be sanctioned to put the pilot training facilities at the College to optimum use. The pattern of the future was beginning to emerge. Graduate Entry (GE) pilots would do their Officer and Basic Flying Training at Cranwell but their Advanced Flying Training at other schools. Cranwell might well become the main one of two or three Basic Flying Training Schools. It was however agreed that the professional training courses for Engineer, Supply and Secretarial officers should continue at Cranwell. Later in 1971, the Air Force

Board discussed with approval the move of the College of Air Warfare from Manby to Cranwell. This would be in line with the philosophy of the College as a 'centre of excellence', an example of professional approach in aviation to newly joined aircrew officers, and would bring a much needed General Duties element to the College.

On 1 March 1971 the flight cadets were so reduced in numbers that Nos 3 and 4 Squadrons were merged to form D Squadron. As can be seen from the details in Annex VI a mixture of passing out parades took place. In the early months of 1971, 1972 and 1973 the last of the flight cadet entries graduated from the College. The passing out parade of the No 101 Entry on 16 March 1973, when the Chief of the Air Staff was the Reviewing Officer, was the last of the old-style flight cadet entry parades. But at regular three or four monthly intervals there were passing out ceremonies for the new Graduate Entry courses. These were of course much smaller in number and often there was a combined exodus as when No1/70 GE Pilots passed out in company with No 2/70 Supply and Secretarial Officers. It was on this parade on 20 August 1971 that Flight Lieutenant The Prince of Wales in the presence of his father HRH Prince Philip, Duke of Edinburgh, received his wings from Air Chief Marshal Sir Denis Spotswood, Chief of the Air Staff.

There is a detailed and admirable account of the royal pupil's flying training at the College in a small privately printed book called '*Wings for the Prince of Wales*' by Squadron Leader R E Johns and published in 1973. Johns was the QFI selected for the responsible instructional duties. Exercise Golden Eagle at Cranwell between 8 March and 30 July 1971 was carried out by a hand-picked team of Qualified Flying Instructors, ground crew and air traffic control and radar operators.

The Prince was no stranger to flying since he had received his Private Pilot's Licence after flying Chipmunks in 1969. From October 1969 to February 1971 he flew the Basset CC1, went solo on it and passed his instrument grading test. But jet flying was of a different order of difficulty and danger. At Annex XIV is given a list of the officers and airmen who were principally involved in the Prince's training.

All were imbued with the caveat that was contained in the operation order from HQ Training Command: 'You are to treat as absolutely overriding the requirement to ensure HRH's safety throughout all aspects and phases of his flying training.' The Prince successfully completed his ground and air handling tests including instrument and night flying and throughout all phases of training satisfied his examiners. Indeed his CFI's final report on the final Handling Test on 8 July 1971 was 'An Above Average performance'.

This adaptability and willingness to learn was put to even further tests when in some crowded days the Prince of Wales flew in a Phantom from RAF Leuchars, in a Nimrod from RAF Kinloss and a Vulcan from RAF Waddington. Finally, on 28 July 1971 the Prince parachuted from an Andover into the English Channel—his last taste of RAF simulation and activity. When he was dined out on 31 July 1971 he was presented with a sculptured model of a Golden Eagle to commemorate his training with the Royal Air Force. The Prince's departure from the College after the parade on 20 August 1971 is described in moving terms by Squadron Leader Johns.

> After a formal luncheon in the College, the Duke of Edinburgh and Prince Charles entered their car at the foot of the College steps with the band playing 'Will Ye No Come Back Again' followed by 'A Life on the Ocean Wave'. At South Airfield nine QFIs flew overhead in an impeccable 'C' formation as the final salute of the Royal Air Force College.

On 14 November 1971 a portrait of HRH The Prince of Wales wearing flying kit painted by Mrs Mara McGregor was presented by the Old Cranwellian Association to the College and unveiled in the Dining Hall by MRAF Sir Dermot Boyle. A new kind of Old Cranwellian was fittingly honoured by the Patron of the Association.[4]

The Prince of Wales returned to Cranwell for refresher flying on JP5s from 23 to 27 October 1972 and for further refresher training from 7 to 11 February 1977.

During this memorable year certain other events were noteworthy. For the first time in the history of the College women officers were on parade as student officers and the Sash of Merit was awarded on parade on 20 August 1971 to Pilot Officer S L Cooke, WRAF. The Headquarters of the University Air Squadrons took up residence at Cranwell, the HQ and the 16 UASs coming under the control of the Air Officer Commanding and Commandant of the RAF College. Other changes included the disappearance of the name Basic Studies Wing. This became Basic Science Wing under Air Commodore C S Betts CBE MA and the Humanities Wing under Air Commodore A Maisner CBE AFC. Both Wings remained located in Whittle Hall.

It must be recorded that the mixing of flight cadet and officer student entries was not entirely without friction and despite the irrevocable changes it implied, the instructional staff looked forward to an ending of the transitional period at Cranwell.

The Future of Cranwell appeared once more on the agenda of the Air Force Board in 1972 and a new, firmer note was sounded. The maximum flying task per year at Cranwell was determined. This inevitably had to include other than university cadet entrants to make up numbers. Two other Basic Flying Training Schools at Church Fenton and Linton-on-Ouse were to be closely associated with Cranwell in basic flying training. Leeming and Valley were to carry out phases of Advanced Flying Training, while Tern Hill was to provide AFT for helicopter pilots. Leeming, Cranwell and Linton-on-Ouse were to be used for refresher flying.

A conference at Cranwell on 25 July 1972 met to discuss the problems arising from the arrival of the College of Air Warfare and other units at Cranwell.[5]

The content of the officer and GST syllabus for the university graduate entries at Cranwell was radically different from that of the former flight cadets. The Introductory and Summer Vacation courses had made the graduates familiar with RAF life and Cranwell routine. For pilots, flying with the University Air Squadron commanded by regular RAF officers had also removed some of the tension if also some of the romance with which younger Cranwell entrants had often approached the flight lines. After leaving their university, cadet graduates had to attend an initial officer and general service training course before passing on to basic flying training. This practice was in parallel with the course carried out at the Officer Cadet Training Unit at RAF Henlow where Direct Entrant graduates, non-graduates and ex-airmen on selection for commissions received their initial officer training. The Graduate Entry at Cranwell therefore was at first treated as a select and separate form of officer training. The old elitism so carefully fostered in College Hall throughout the years was still meaningful. But the numbers concerned were extremely small. The length of time at Cranwell was barely a twelve-month as may be seen from Annex X. Furthermore the students were older, more mature, and sometimes with quite unaccustomed appendages like wives and family responsibilities. The civilian staff and the Cranwell instructors had to learn new approaches. These were no longer rather brash or rather diffident boys straight from school.

But where flying was concerned there was a great levelling and it was in handling the Jet Provost that the graduates met their greatest intellectual as well as character challenge. This after all for aircrew was the object of the exercise. Not that officer qualities were not tested in the many activities for which GST at Cranwell has always been noted. Survival exercises in Snowdonia and other testing localities, hill walking, canoeing, rock climbing and orienteering all figure in the Operations Record Book reports of training.

Although traditional activities and links, especially with local bodies, were maintained the changed character and the numerical strength of the College produced some unusual results. For example, the visiting cadets from the USAF Academy at Colorado Springs beat the College XV at its own game of rugby.

In Trenchard Hall a great variety of Engineering courses were being carried out. Among them in 1973 were the following:

Initial Engineering Courses for Graduate Entrants and Direct Entrants.
Staff Officers Computer Courses.
Senior Officers Computer Appreciation Courses.
Aerosystems Engineering Courses.
Advanced Maintenance Engineering.
Engineering (Overseas officers) Courses.
Electrical and Instruments Courses (Overseas officers).

Professional training courses for Supply and Secretarial Officers including the Graduate Entrants were also being conducted at Cranwell. There was thus an increasing element of ground training at the College. Three officers' messes were active—College Hall, York House and Trenchard Hall (formerly No 2) Mess—and the tendency towards separatism or local loyalty was one which successive Commandants had to watch with care.

On 2 January 1974 the Department of Air Warfare formed at Cranwell with Air Commodore J R Rogers CBE as Assistant Commandant. This proudly autonomous unit now became part of the College. Its Dominie Squadron had arrived in the last month of 1973 to become part of Flying Wing. A whole new range of courses on Air Warfare, Armament, Weapons Employment, Electronic Warfare and other specialist aviation courses brought an older, more experienced type of student from an international range of air forces into the College environment. Flying certainly received a boost. Ground trials carried out on Harrier aircraft in January-February 1975 seemed quite appropriate to the South airfield at Cranwell. On 16 October 1975 it is recorded that the new Hawk trainer made its maiden appearance at Cranwell. The flying activities of the Graduate Entrants were also streamlined. A basic 75-hour course on the Jet Provost Mk 5A normally led to the award of wings. The 75 hours were obtained from 28 weeks broken down into:

$2\frac{1}{2}$ weeks ground school.

$10\frac{1}{2}$ weeks half time flying/ground school.

2 weeks leave.

13 weeks full time flying.

In addition pilots had to put in 25 hours on the Jet Provost Instrument Trainer. This basic flying training was to be followed by 60 hours of advanced training on the JP5A Group 1 Phase 1 course. Most GD pilots therefore were spending approximately one year under the influence of Cranwell.

The links with tradition seemed strong when HM The Queen presented the new Queen's Colour to the College on 30 May 1975. Sixteen Jet Provosts flew past Her Majesty in diamond formation as she planted a commemorative tree.

Cranwell, like the rest of the Royal Air Force, felt the impact of the Redundancy Scheme. In the compulsory zone serving at Cranwell were 1 group captain, 17 wing commanders and 20 squadron leaders. Wings ceremonies now began to take place inside College Hall and at the dining-in night for 15 pilots of No 18 GE Course, prizes were presented before dinner. Eventually a decision was taken to concentrate a series of small fragmented ceremonies and on 7 May 1976 Air Marshal R D Roe CB AFC, Air Officer Commanding-in-Chief Training Command reviewed the first Annual Prizes Parade on the College Parade Ground.

In 1976 also, on 12 April, the Commandant of the Central Flying School, Air Commodore J M D Sutton, his Headquarters Staff, Examining Wing and the Jet Provost element and ground school of the CFS arrived at Cranwell. The event was marked by a flypast of 16 JPs and the Vintage Pair, Meteor and Vampire aircraft. The Central Flying School remained, not entirely happily, at Cranwell until late 1977 when the Training Flight moved to RAF Leeming. CFS was dined out from York House Officers' Mess on 22 November 1977 and a tree was planted and a plaque installed on the north-east corner of the large grassed area opposite the Station HQ Building 259, to commemorate the stay at Cranwell of CFS. A Station Commander had indeed reappeared at Cranwell when the old title of OC College Unit was abolished on 9 July 1975.

The Organization of the Base

Cranwell, as in both World Wars, had now become rather complicated. In November 1975, before the arrival of CFS, there were three departments—Air Warfare, Specialist Ground Training and Officer and Flying Training.[6] The presence of overseas students from old and new Commonwealth air forces, from the air forces of NATO allies and foreign air forces was again a strong link with tradition at Cranwell.

The performances of the Poachers, the CFS Vintage Pair and the solo Jet Provost did much to bring flying before the general public and to the attention of all who worked at or visited Cranwell. The Poachers formed in 1963 were led in the early years of the decade by Squadron Leader J B Robinson AFC who was a flying instructor to HRH The Prince of Wales in 1972. The team of four JP Mk5 aircraft added to the reputation of the College with displays throughout the British Isles and on the Continent. It was fitting that their normal concluding manoeuvre should be 'The Prince of Wales break' tracing feathers in the sky and ending in a thrilling crossover and twinkle rolling exit.

The first conference of the newly formed Administrative Branch was held in Whittle Hall on 15 April and in May General R E A Glavany, the Commander-in-Chief of the French Air Force, came to be briefed on the training methods of the RAF College and of Central Flying School. Cranwell was well represented at Finningley in July when the 25th Anniversary of the Queen's Coronation was celebrated. A box of four Cranwell Dominies took part and Cranwell Jet Provost aircraft formed the figure 5 of the symbolic '25' led by Wing Commander John Jewiss, the Chief Flying Instructor of the RAF College. A static display was also mounted to depict the roles of the College—officer training, specialist ground and advanced air studies.

Mr J S Robinson retired from the Station Post Office after 22 years of service.[7] His father J W Robinson had opened the old Post Office in 1916 and it is very pleasant to note that the new owner of the shop though not related also has the name of Robinson.

On 30 September 1977 the Dominie Squadron disbanded and aircraft and pilots were absorbed into the Dominie Squadron at RAF Finningley. As the year drew to a close CFS departed for Leeming and the College like other RAF units became involved in the manpower crisis caused by detachment of personnel to take the place of striking civilian firemen on fire-fighting duties. A long period of austerity and self-service ensued and College Hall became accustomed to the uncommon sight of staff officers, students and visitors trooping past the hot plates in the kitchens to collect their meals.

There is a point at which history becomes indistinguishable from a current affairs commentary but the year 1978 will bear examination for one or two events of importance. The Old Cranwellian Association Annual Dinner was held on 17 June in the College Hall Officers' Mess. The principal guest was HRH The Prince of Wales. With the many changes in conditions of officer recruitment and service, serious thought was being given to the question of who was entitled to be called an 'Old Cranwellian'. In the event after much discussion and canvassing of opinion among former cadets of all ages, a decision was reached in 1980 that those who had received their training at the RAF College leading to a permanent commission in the Royal Air Force should be eligible for membership of the Old Cranwellian Association.[8] Thus the Association will continue and flourish and not perish with the demise of the last of a former pattern of training. Cranwell preeminently has always been a living institution.

As if to demonstrate the qualities of enterprise and courage typical of a new generation, a member of No 29 Graduate Entry set off in a single-engined aircraft to fly round the world. This was Flying Officer J P Warren-Wilson, an Oxford graduate, who accompanied by a university colleague, Mr N Walsh, a civil engineer, took off from the South airfield at 1125 hours BST on 31 March 1978. The flight was approved under the RAF Expedition Training Scheme and received some backing from public funds. The spirit of the 1930s was indeed still active. Warren-Wilson flew back from Reykjavik to Cranwell and arrived on 26 July 1978 in poor visibility and light rain to a welcome from the Commandant and in front of BBC and ITV cameras.

Though such things as Lords Taverners' Day continued as usual (now with the force of tradition rather than one-off innovation) there were many changes taking place. The changes at Cranwell such as the departure of CFS for Leeming stemmed from changes of thinking at Air Force Board level about the future of the RAF College. As early as 1975 the working of the Graduate Entry scheme and the expensive coexistence of two separate centres of Initial Officer Training at Cranwell and Henlow had led the Air Member for Personnel to set up a Working Party to carry out a root and branch examination of the depth, content and timing of Initial Officer Training in the Royal Air Force. This

Working Party chaired by Air Commodore R K Hooks who was Director of Training (Ground) (RAF) submitted its Report to the Air Force Board in June 1976. The main recommendations of the Report were that all branches of the Royal Air Force should be accorded the same Initial Officer Training (IOT). This should be given as a single package as soon as possible after entry into the Service. All Initial Officer Training should be concentrated at Cranwell.

Thus the concept of the Single Gate Entry Initial Officer Training came into being. The Report also laid stress on the importance of strengthening the leadership and motivational aspects of IOT.

The Air Force Board Standing Committee in its Conclusions on 7 April 1977 recognized the steps necessary to accomplish the Single Gate Entry for all officers in the Royal Air Force, the closure of the Officer Cadet Training Unit at Henlow and the inclusion of Direct Entrants into General Duties professional pilot training courses at Cranwell. These steps were:

a. the establishment of all IOT at Cranwell;

b. the rationalization of basic flying training between Linton-on-Ouse and Cranwell.

c. the concentration of all navigator training at Finningley.

d. the move of CFS as an independent unit to Leeming.

In a comprehensive note from the Air Members for Personnel and Supply and Organization in June 1977, the Air Force Board accepted that the Working Party on IOT had found no difference in the quality of the end product from Cranwell and Henlow so far as performance in early appointments was concerned and there was no evidence that the graduate should be given a special form of IOT. Merging the courses at Henlow and Cranwell would lead to better *esprit de corps*, and by having the combined course at Cranwell, all officers would benefit from the long established Cranwell environment and tradition.

It was clear that the role of the RAF College was itself being modified and its redefinition was along the lines 'The centre of all Initial Officer and as much Initial Specialist Training and associated advanced training as possible'. It was also clear that the concentration of IOT at Cranwell and the proposed rationalization of flying training tasks would result in very considerable savings in manpower and running costs.

A letter from Director of Organization and Administrative Plans (RAF) on 12 July 1977 to HQ RAF Support Command issued the instruction that in the future all Initial Officer Training would be conducted at Cranwell. It was proposed to achieve this by the end of 1979.

Course No 34 which entered Cranwell on 4 December 1978 was known as IOT 34 Course and the name Department of Initial Officer Training came into use.[9] The composition of succeeding courses changed considerably. No 37 IOT Course, for example, was the first at the RAF College to contain WRAF cadets (as opposed to student officers). The Course comprised 52 Direct Entrants and 34 former serving airmen and airwomen, of whom nine were WRAF cadets. The epilogue to this history, Cranwell Today, gives a very brief picture of the present situation at the RAF College. Cranwell in 1980 has changed in very many ways from the station it was in 1920 and changes will continue. The Cranwell badge with its motto 'Superna Petimus' remains as a challenge to the training of the future leaders of the Royal Air Force.

14 *Cranwell Today*

At the time of writing—December 1980—there are four departments at the RAF College, each directed by an officer of group captain rank. In West Camp, using Whittle Hall for instructional and office accommodation, and the 234 single officers' quarters of College Hall for cadets and staff, is the Department of Initial Officer Training. Headquarters University Air Squadrons is also accommodated in College Hall. In East Camp, Trenchard Hall houses the Department of Air Warfare and the Department of Specialist Ground Training. The Commandant is an air vice-marshal and has an air commodore as Deputy Commandant. A group captain commands the Basic Flying Training School, a lodger unit (responsible functionally direct to Headquarters Royal Air Force Support Command) which provides all flying and engineering services at Cranwell. There is no Station Commander as such, the administrative backing to the College and the Basic Flying Training School being provided by an Officer Commanding Support Unit who is a wing commander.

The student population of the Department of Initial Officer Training is mainly composed of:

University Cadet Graduates	(UCG)
Graduate Direct Entrants	(GDE)
Non-Graduate Direct Entrants	(NGDE)
Ex-Serving Airmen	(Ex-SA)

They may be men or women and are trained to a common syllabus on an 18-week officer training course. After graduation and commissioning from this course officers go separate ways for specialist training according to their branches. In addition, professionally qualified doctors, dentists, nurses, padres, legal officers, officers of the Royal Auxiliary Air Force and Royal Observer Corps, officers re-entering the Service, and officers of the Royal Navy and Army entering the Royal Air Force undergo a four-week course. The Department has expanded greatly to undertake these tasks; the staff now number over 140 and some 450 students might be in residence at any one time. Domestic accommodation and messing facilities are widely dispersed. During the cadets' first six weeks of the course they live in East Camp where they sleep in converted barrack blocks; for their second six weeks they move to the former Junior Flight Cadets' Mess in West Camp; finally, for their last six weeks, as the Senior Course, they live in College Hall Officers' Mess amidst the traditions of past years. In 1980 some 600 officers were commissioned from Cranwell.

Competition is as strong as ever for Sovereign Squadron, and the Knocker, Chimay and Ferris Trophies. Therefore well-known activities are preserved which have traditionally played such an important part in stimulating competitive spirit among cadets.

The first year of Single Gate Initial Officer Training produced excellent sports results, perhaps the most notable being in the inter-collegiate Winter Games in which Dartmouth was eclipsed by six games to nil and Sandhurst thoroughly beaten by five games and one drawn—this performance is believed to have been unequalled in the past.

Headquarters University Air Squadrons is directly responsible to the Commandant for organizing, controlling and administering the 16 University Air Squadrons, with over 700 members, located throughout the UK. The Squadrons give flying training in Bulldog aircraft to full-time undergraduates studying for a degree at one of the 80 affiliated Universities and Colleges of Higher Education. Members of the Squadrons are either Royal Air Force Volunteer Reservists, who have no obligation to join the RAF, or University Cadets, who are regular RAF officers undertaking full-time degree studies.

Along Cranwell Avenue in East Camp, the Department of Air Warfare continues to provide post-graduate training in air warfare. The emphasis of this training is on air operations, operational requirements and airborne avionic systems, sensors, weapons and the associated tactics. The Air Warfare Course, sponsored by the Vice Chief of the Air Staff and some five months long, is the most senior staff training course in the RAF. This course prepares group captains and wing commanders for appointments concerned with the planning and execution of air operations, while the 50-week General Duties Aerosystems Course trains less senior officers who are destined for employment as systems specialists in operational units, within the Operational Requirements staffs, and in the Research and Development establishments. The Department runs a host of other shorter, highly specialized courses to meet the increasingly complex and technological needs of the Royal Air Force.

Finally in East Camp there is the Department of Specialist Ground Training which has not escaped considerable change in recent months. Engineer officers now undertake up to 13 months initial specialist training at the College to fit them for two new fields of employment—Aerosystems Engineering and Communications-Electronics Engineering. Advanced specialist training, in association with Loughborough University of Technology, is also given to selected Engineer officers. Supply officers attend a course of approximately four months professional training to enable them to fill junior appointments in station supply organizations. Specialist training for Secretarial officers of the Administrative Branch moved from Cranwell to Hereford in July 1980.

But Cranwell also provides the airfield and facilities for the Basic Flying Training School. GD Pilots will attend the standard 36-week basic flying training course at one of the RAF's Basic Flying Training Schools (BFTSs). A shorter course of 30 weeks is available at Cranwell only for students who have accumulated substantial flying experience and can meet certain selection criteria. UCGs who have successfully completed the full UAS flying training syllabus will normally qualify for entry to the 30-week course at Cranwell. GDEs who have completed two years as VR members of a UAS may also qualify for this course. Student pilots undergoing basic flying training at Cranwell will therefore stay for over a year and if selected for fast jet training they will continue at the College for a further 11 weeks to complete the first phase of advanced flying training before moving on to the Hawk course at Valley. Time at Cranwell therefore could be: 18 weeks + 30 weeks + 11 weeks. Even this is much short of the two/two and a half or three years experienced by former flight cadets. However, the training is far more concentrated than in those earlier days.

Officers, including UCG and GDE, who cannot meet the selection criteria for the shortened course may do their basic flying training at Church Fenton or Linton-on-Ouse alongside NGDE or Ex-SA students.

Navigators leave the College after officer training and go to Finningley for their flying training. Pilots selected for multi-engine or helicopter training leave Cranwell after basic flying training and go to an Advanced Flying Training School for about five months leading to the award of the pilots flying badge. Officers of other branches leave Cranwell to complete professional training in their branches at other RAF stations.

The RAF College has become an amalgam of various activities, a large base on which Initial Officer Training forms only one part of the unit task. The role of the Commandant was never more significant as the unifying factor of the College.

Though Cranwell is now the alma mater for all RAF officers entering the Royal Air Force, it is not to say that Cranwell has ceased to be a centre of excellence. Cranwell is also a centre of advanced studies appropriate to a college. The deliberate mixing of officers from different forms of entry has produced a sense of comparison and challenge to attain the highest standards. The need to avoid peaking in course planning and to use aircraft and instructors to the best advantage has also had this side effect. Performance in aviation skill has become a major criterion for professional success.

There can now be one kind of officer only—trained at Cranwell. Perhaps Lord Trenchard would have approved of this imaginative thinking. What might have caused him some concern is the apparently increasing tendency for Cranwell to become a ground training station. Being an officer and an aviator were almost synonymous in an air force that was less sophisticated and smaller. But flying was, and still remains, the aim and object of the exercise. The environment which Trenchard wanted for training young men still obtains at Cranwell. It is in the country, not so remote as formerly but still removed from the vice and pleasure areas of London. Young men and women need to create their own enjoyment, develop their own activities and concentrate on training to become future leaders of the Royal Air Force, the world's first independent military air service. As they go about the College the sight and sound of aircraft must be the background to their training. Moreover, if the stay at Cranwell is to be short, compared with that of former times, the quality of training must be of the highest standard.

This standard will not be easy to maintain when passing out parades occur every six weeks. But the passing out parades are again large and are impressive displays as in the past, and the College was deeply pleased to learn that Her Majesty The Queen had approved the title 'The Queen's Review' for the annual prize-giving ceremony known as the College Review since 1976. After over 60 years, therefore, Cranwell was honoured to have the equivalent of Dartmouth's Lord High Admiral's Parade and Sandhurst's Sovereign's Parade. Most happily Her Majesty The Queen also agreed to visit Cranwell on 24 July 1981, together with the Duke of Edinburgh, for the first Queen's Review. The College is once again filled to capacity, mainly with young, fit men and women. It is living as it did in the days of its fullest occupation in two World Wars, especially from 1939 to 1945, when the base was a veritable United Nations in miniature.

The present occupants of College Hall whose stay is only one third of the 18 weeks course will still be able to say in the future: 'I too was at Cranwell'. Something of the great traditions of the past will become part of their personal heritage.

Annexes

General Memorandum No 14

24 February 1916

Air 3434/1916

1. On and after 1st April 1916 all officers under instruction in aeroplanes, kite balloons, and airships are to graduate as pilots at Cranwell. The procedure for graduation is shown in Appendices I. to IV. herewith.

2. As accommodation becomes available, all officers and men under instruction in aeroplanes, kite balloons, and airships will be sent to Cranwell for a finishing course and for graduation. Commanding Officers of Stations are to note on the "Weekly report of progress of officers under instruction" the names of those recommended for this final course. In addition to this, all officers under the rank of Squadron Commander will be required to re-qualify periodically.

3. The Commodore will be informed from time to time of the number of seaplane pilots required, and the names of the officers recommended by Commanding Officers of Stations for this Service. Officers for further training as seaplane pilots will be selected from the list of those recommended.

4. After graduating at Cranwell as aeroplane pilots, officers will, as a general rule, be appointed to Freiston for about 14 days practical work in gunnery and bomb-dropping.

5. On the completion of this course, officers will be appointed to Home Stations; they will not be appointed for Service abroad immediately if this can be avoided. This will enable officers to gain further experience.

Officers selected to qualify as seaplane pilots will be appointed to seaplane stations for necessary training.

6. Officers who show lack of zeal and insufficient progress in their instructions, or who, in the opinion of the Commanding Officer of the Training Establishment, are not likely to become good pilots, will be liable to discharge from the Service.

7. A supply of graduation certificates will be made to all concerned as soon as possible in the form shown in Appendix V.

From RNAS Memoranda,
General & Temporary 1916

Extracts from the log book of Assistant Paymaster J H Lidderdale, RN, CFS, Upavon

8 Oct 1912	Commenced instruction under Lt A M Longmore, RN in Maurice Farman biplane. Had 3 lessons of about 20 mins each with Lt Longmore up to 20 Dec.
13 Jan 1913	Made first solo flight.
14 Jan 1913	In the afternoon made 5 landings at a fixed spot. 3 were within correct distance of mark. Then went for certificate and obtained it. During figures engine failed four times due probably to water in petrol. On each occasion closed throttle and descended and opened throttle again almost immediately. Engine at once recovered. [Lidderdale logged 160 hours before his first flight at Cranwell.]
20 Feb 1916 Cranwell	Maurice Farman 3008 with Air Mechanic Swales as passenger.

Last entry in log book was

17 Aug 1918 Cranwell	DH6 (C2128) Colonel Smith as passenger. 25 minutes over Scopwick and vicinity.

Lieutenant-Colonel (Wing Commander) J H Lidderdale RN (RAF) died at a very early age and there is little doubt that years of heavy work under pressure were largely responsible for his early demise. There is a collection of Lt Lidderdale's photographs and papers in the RAF Museum. Below is an extract from a letter by AVM F H Maynard to Gp Capt D C S Evill on the occasion of Lidderdale's death:

> I first met him in December 1915 when he was sent to help Sir Godfrey Payne [sic] to get Cranwell going. He and myself (then a very junior and inexperienced Flying Officer) were the only Executive officers for some time and, since Cranwell was nothing but a farm when we arrived, there was much to be done. Lidderdale was by nature, as you no doubt know, both a terrific worker and a borne [sic] organizer. In those early months at Cranwell he appeared never to stop work. Cranwell was in fact Lidderdale and Lidderdale Cranwell in those early months. . . He knew every airman by name up to the middle of 1916... He remained at Cranwell throughout 1917 and so far as I can remember (I was not there myself) carried on as First Lieutenant until the RAF was formed on 1/4/18. He shortly after was moved to the staff of No 12 Group formed with headquarters at Cranwell

34 Aero engine display room, 1966

35 Cadets from USAF Academy, Colorado Springs, with RAF Cadets at drill, 1960

36 Line up of Jet Provosts, 1966

37 Main workshops, Trenchard Hall, 1966

38 **Chipmunk primary trainers over College Hall, 1953**

39 Jet Provost basic trainer, 1966

40 Jet Provost

41 North American Harvard T2 advanced trainer, 1950

42 Instruction on a Rover gas turbine engine, 1965

43 View of the College, 1961

44 HM The Queen plants a tree, July 1960

45 After the Parade, July 1960

46 RAF Swimming Championships in the College swimming pool, 1962

47 Whittle Hall, 1966

48 Opening ceremony, Trenchard Hall, May 1966

49 External view of main College gymnasium and swimming pool, 1966

and constituting part of Inland Area, RAF. He was, I believe, General Scarlett's right hand man on this Group staff and he stayed there until the end of the War.

Early in 1919 he was posted to the Air Ministry to help in the demobilization of the RFC and RNAS. The next difficult problem he was given was the revision of King's Regulations. This was of course a terrific task but Lidderdale tackled and all but completed it before his health gave way...

Air Ministry, 17 Nov 1933.

ANNEX III

Record of Balloons and Airships at Cranwell, 1916-1918[1]

Balloons, spherical 1916-1917

COAL-GAS	HYDROGEN
North Star	Achilles
Shark	Velocity
Shrimp	
Meteor	
Plover	
Leda	
Lapwing	

Airships

Rigid No 9	October 1917. Damaged at Howden and retained there
Rigid No 25	April 1917 to September 1919
North Sea 4	October 1918. Flew from Kingsworth via Cranwell to Howden and East Fortune
North Sea 11	April to June 1918
Parseval No 6	August 1917 to January 1918
C Star 8	July 1918
Coastal 13	January 1916 to October 1917 (Considerable trouble with her engines and ground mishaps)
SS 28a (SS28)	January 1918 to 23 May 1919 (Deflated)
SS 29	Made a forced landing at Rauceby on her way to Cranwell from Folkestone on 3 January 1918. Deflated 27 January 1919
SS 30a	December 1917 to January 1919
SS 31	Flew in December 1916 to Cranwell from Bridgenorth. On 11 July 1917 collided with roof of the rigid shed, carried away her under carriage and broke her propellers. Made a free balloon landing some three miles from station
SS 31a	Rebuilt from SS 31—1918 to January 1919

[1] Cranwell was much used as a stopping place on acceptance flights of airships from Wormwood Scrubs to Howden in Yorkshire.

SS 37	Flew from Pembroke to Cranwell in May 1917. Rebuilt at Cranwell August 1917. Deflated 15 May 1919
SS 39	Wrecked near Sleaford in November 1916 on way from Wormwood Scrubs to Cranwell. In May 1917 made a forced landing when rigging got caught in some trees. On 20 July 1917 fatal accident to officer and men. Rebuilt at Cranwell August 1917. Deflated January 1919
SSP 1	March 1918 to January 1919
SSP 5	March 1918 to January 1919
SSP 6	April 1918 to February 1919
SSZ 23	January to March 1918
SSZ 60	April 1918 only
SSZ 61	May 1918 to 2 January 1919

Commandants RAF Cranwell[1]

1920 – 1923 Air Commodore C A H Longcroft CMG DSO AFC
1923 – 1926 Air Commodore A E Borton CB CMG DSO AFC
1926 – 1929 AVM F C Halahan (AVM 1928) CMG CBE DSO MVO
1929 – 1933 AVM A M Longmore (AVM 1930) CB DSO
1933 – 1934 AVM W G S Mitchell CBE DSO MC AFC
1934 – 1936 AVM H M Cave-Brown-Cave (AVM 1935) DSO DFC
1936 – 1938 AVM J E Baldwin DSO OBE

See Chapters 7 to 9 for the period 1938-1946

1946 – 1949 Air Commodore R L R Atcherley CBE AFC
1949 – 1950 AVM G R Beamish CB CBE
1950 – 1952 AVM L F Sinclair GC CB CBE DSO
1952 – 1956 Air Commodore H Eeles CBE
1956 – 1958 Air Commodore T A B Parselle CBE
1958 – 1961 Air Commodore D F Spotswood CBE DSO DFC ADC
1961 – 1963 Air Commodore E D Mac K Nelson CB
1963 – 1964 Air Commodore M D Lyne AFC
1964 – 1967 AVM I D N Lawson CB CBE DFC

Merger with RAF Technical College Henlow 1965-1966

1967 – 1970 AVM T N Stack CVO CBE AFC
1970 – 1972 AVM F D Hughes CBE DSO DFC AFC MA

No. 1 Graduate Entry Course began 1 September 1970.

1972 – 1975 AVM R D Austen Smith DFC

The RAF Flying College Manby closed and the Department of Air Warfare was opened at Cranwell in January 1974.

1975 – 1978 AVM W E Colahan CBE DFC

RAF Central Flying School was based at Cranwell between April 1976 and November 1977.

1978 – 1980 AVM D Harcourt Smith DFC
1980 – 1982 AVM B Brownlow OBE AFC
1982 – AVM R C F Peirse

[1] The decorations of the distinguished officers named were those they held at the time of their holding the office of Commandant.

Commandants RAF Technical College Henlow[1]

1949–1952	Air Commodore D W R Ryley CBE
1952	Air Commodore J G Franks CBE
1952–1954	Air Commodore H D Spreckley CB OBE
1954–1956	Air Commodore C L Dann CBE BSc
1957–1959	Air Commodore T U C Shirley CBE MIEE AFRAeS
1959–1961	Air Commodore N C S Rutter CBE MIMechE FRAeS
1961–1963	Air Commodore J C Pope CBE MIMechE AFRAeS
1963–1966	Air Commodore E M T Howell CBE FRAeS

[1] The decorations of the distinguished officers named were those they held at the time of their holding the office of Commandant.

Assistant Commandants RAF Cranwell

1920 – 1921	Wing Commander A L Godman CMG DSO
1921 – 1923	Wing Commander A S Barratt CMG MC
1923 – 1924	Wing Commander L W B Rees VC OBE MC AFC
1924 – 1929	Wing Commander E M Murray DSO MC
1929 – 1931	Wing Commander D C S Evill DSC AFC
1931 – 1935	Wing Commander P Babington MC AFC
1935 – 1937	Group Captain J H Herring DSO MC
1937 – 1938	Wing Commander C N Lowe MC DFC
1938	Wing Commander E J Kingston – McCloughrey DSO DFC
1938 – 1939	Group Captain R Halley DFC AFC

See Chapters 7 to 9 for the period 1939-1946

1947 – 1948	Group Captain J R A Peel DSO DFC
1948 – 1951	Group Captain J O W Oliver CB DSO DFC
1951 – 1952	Group Captain W H Kyle CBE DSO DFC ADC
1952 – 1954	Group Captain E D Mac K Nelson CB ADC
1954 – 1957	Group Captain D H Lee DFC
1957 – 1959	Group Captain H N G Wheeler CBE DSO DFC AFC ADC
1959 – 1960	Group Captain L M Hodges CBE DSO DFC
1960 – 1963	Group Captain T P Seymour CBE
1963 – 1964	Group Captain A C Deere DSO OBE DFC ADC
1964 – 1965	Group Captain G C Lamb AFC

Merger with RAF Technical College Henlow took place in 1965 – 1966.

1965 – 1966	Air Commodore N Cameron DSO DFC (Department of Cadets)
1965 – 1968	Air Commodore J S Rowlands GC OBE BSc (Department of Engineering)
1966 – 1968	Air Commodore R G Wakeford MVO AFC (Department of Cadets)
1968 – 1971	Air Commodore C S Thomas CBE (Department of Engineering)
1968 – 1971	Air Commodore J E Bazalgette DFC (Department of Cadets which on 15 June 1970 became Department of Training)

1971 – 1972	Air Commodore C S Betts CBE MA (Department of Engineering)
1971 – 1973	Air Commodore A Maisner CBE AFC (Department of Training)
1972 – 1975	Air Commodore W C Taylor OBE CENG AFRAeS (Department of Engineering which became Department of Specialist Ground Training)
1973 – 1974	Air Commodore B Brownlow OBE AFC (Department of Officer and Flying Training)
1974 – 1976	Air Commodore J R Rogers CBE (Department of Air Warfare) wef 2 January 1974
1974 – 1977	Air Commodore M M J Robinson (Department of Officer and Flying Training)
1976 – 1978	Air Commodore B W Plenderleith MA CENG MRAeS MBIM (Department of Air Warfare)
1975 – 1978	Air Commodore H F Glover CENG MIMechE (Department of Specialist Ground Training)
1977 – 1979	Air Commodore H S Carver MVO[1] (Department of Officer and Flying Training)
1979 –	Air Commodore E G P Jeffery (Deputy Commandant)

[1] Air Commodore H S Carver became the Deputy Commandant of the RAF College when the formation was split into three departments: Department of Air Warfare, Department of Initial Officer Training, Department of Specialist Ground Training. The departments have a Director in the rank of group captain.

Reviewing and Inspecting Officers RAF Cranwell

20 Dec 1920	The Rt Hon Winston Churchill MP (Secretary of State for War and Air)
	Air Marshal Sir Hugh Trenchard Bart KCB DSO (Chief of the Air Staff)
19 Dec 1921	AVM Sir John Salmond KCB CMG CVO DSO (substituting for Captain the Rt Hon F E Guest)
15 Aug 1922	Captain the Rt Hon F E Guest CBE DSO MP (Secretary of State for Air)
	Air Chief Marshal Sir Hugh Trenchard Bart KCB DSO ADC (Chief of the Air Staff)
20 Dec 1922	Air Chief Marshal Sir Hugh Trenchard Bart KCB DSO ADC
14 Aug 1923	AVM Sir Geoffrey Salmond KCMG CB DSO
18 Dec 1923	Air Commodore J M Steel CMG CBE (in the absence of CAS)
30 Jul 1924	Field Marshal Sir William Robertson Bart GCB GCMG KCVO DSO
17 Dec 1924	AVM Sir Philip Game KCB DSO
28 Jul 1925	Sir Philip Sassoon Bart CBE CMG MP (Under Secretary of State for Air)
15 Dec 1925	AVM T I Webb-Bowen CB CMG
29 Jul 1926	Sir Philip Sassoon Bart CBE CMG MP
10 Dec 1926	Air Chief Marshal Sir Hugh Trenchard Bart GCB DSO
29 Jul 1927	The Rt Hon Sir Samuel Hoare Bart GBE CMG MP (Secretary of State for Air)
16 Dec 1927	Air Marshal Sir John Salmond KCB CMG CVO DSO ADC
27 Jul 1928	AVM Sir Philip Game KCB DSO
14 Dec 1928	MRAF Sir Hugh Trenchard Bart GCB DSO DCL LLD
26 Jul 1929	AVM Sir Edward Ellington KCB CMG CBE
13 Dec 1929	AVM Charles L Lambe CB CMG DSO
25 Jul 1930	AVM E R Ludlow-Hewitt CB CMG DSO MC
19 Dec 1930	Air Chief Marshal Sir John Salmond KCB CMG CVO DSO LLD
24 Jul 1931	AVM Sir Charles L Lambe KCB CMG DSO
18 Dec 1931	AVM T I Webb-Bowen CB CMG
22 Jul 1932	Air Chief Marshal Sir John Salmond KCB CMG CVO DSO LLD
16 Dec 1932	AVM R H Clark-Hall CMG DSO (AOC Coastal Area)

14 Jul	1933	Air Marshal Sir Robert Brooke-Popham KCB CMG DSO AFC
15 Dec	1933	MRAF Lord Trenchard GCB DSO DCL LLD
27 Jul	1934	Air Marshal Sir Hugh Dowding GCVO KCB CMG
14 Dec	1934	Air Chief Marshal Sir Edward Ellington KCB CMG CBE
26 Jul	1935	Air Marshal Sir Cyril Newall KCB CMG CBE AM
13 Dec	1935	Air Marshal Sir Arthur M Longmore KCB DSO
31 Jul	1936	MRAF The Viscount Trenchard GCB GCVO DSO DCL LLD
18 Dec	1936	Air Marshal Sir Hugh Dowding GCVO KCB CMG
30 Jul	1937	Field Marshal Sir Cyril Deverell GCB KBE (Chief of the Imperial General Staff)
17 Dec	1937	Air Chief Marshal Sir Cyril Newall KCB CMG CBE AM (Chief of the Air Staff)
29 Jul	1938	MRAF Sir John Salmond GCB CMG CVO DSO DCL LLD
16 Dec	1938	Air Marshal Sir Arthur M Longmore KCB DSO
28 Jul	1939	General Lord Gort VC KCB CBE DSO MVO (Chief of the Imperial General Staff)
7 Apr	1948	MRAF Lord Newall GCB OM GCMG CBE AM
8 Apr	1949	*Air Chief Marshal His Royal Highness, The Duke of Gloucester*
27 Jul	1949	MRAF The Viscount Trenchard GCB GCVO DSO DCL LLD
14 Dec	1949	Field Marshal Sir William Slim GBE KCB DSO MC
5 Apr	1950	MRAF Lord Tedder of Glenguin GCB DCL LLD
26 Jul	1950	MRAF Sir John Slessor GCB DSO MC
13 Dec	1950	MRAF The Viscount Portal of Hungerford KG GCB OM DSO MC DCL LLD
11 Apr	1951	Admiral of the Fleet Lord Fraser of North Cape GCB KBE
1 Aug	1951	*Her Royal Highness The Princess Elizabeth, Duchess of Edinburgh*
12 Dec	1951	General Sir Gerald Templer KCB KBE CMG DSO
9 Apr	1952	The Rt Hon The Lord de l'Isle and Dudley VC (Secretary of State for Air)
30 Jul	1952	Field Marshal The Rt Hon Earl Alexander of Tunis KG GCB GCMG CSI DSO MC DCL LLD
11 Dec	1952	MRAF Sir John Slessor GCB DSO MC
14 Apr	1953	Air Chief Marshal Sir Hugh P Lloyd KCB KBE MC DFC LLD
28 Jul	1953	*His Royal Highness The Duke of Edinburgh*
15 Dec	1953	Air Chief Marshal Sir William F Dickson GCB KBE DSO AFC
6 Apr	1954	Field Marshal Sir John Harding GCB CBE DSO MC (Chief of the Imperial General Staff)
27 Jul	1954	General Alfred M Gruenther (SACEUR)
14 Dec	1954	Admiral of the Fleet Sir Rhoderick R McGrigor GCB DSO LLD
5 Apr	1955	Air Chief Marshal Sir Arthur P M Sanders GCB KBE ADC

13 Dec	1955	Air Chief Marshal Sir Francis J Fogarty KCB KBE DFC AFC
10 Apr	1956	Air Chief Marshal Sir Dermot Boyle KCVO KBE CB AFC
31 Jul	1956	Field Marshal The Viscount Montgomery of Alamein KG GCB DSO
9 Apr	1957	The Rt Hon George Ward MP (Secretary of State for Air)
30 Jul	1957	Admiral of the Fleet The Earl Mountbatten of Burma KG PC GCB GCSI GCIE GCVO DSO LLD DCL DSc
17 Dec	1957	Air Chief Marshal Sir Ronald Ivelaw-Chapman GCB KBE DFC AFC
1 Apr	1958	Air Chief Marshal Sir George H Mills KCB DFC ADC
29 Jul	1958	General Lauris Norstad (SACEUR)
16 Dec	1958	Air Marshal Sir Richard L R Atcherley KBE CB AFC
28 Jul	1959	Air Chief Marshal Sir Harry Broadhurst KCB KBE DSO DFC AFC
15 Dec	1959	General Sir Francis Festing GCB KBE DSO ADC (Chief of the Imperial General Staff)
25 Jul	1960	*Her Majesty Queen Elizabeth II*
13 Dec	1960	Admiral Sir Caspar John GCB
1 Aug	1961	General D'Armee Aerienne P M V Stehlin CVO OBE (Chief of the Air Staff, French Air Force)
19 Dec	1961	Air Chief Marshal The Earl of Bandon GBE CB CVO DSO
31 Jul	1962	MRAF Sir Thomas G Pike GCB CBE DFC (Chief of the Air Staff)
18 Dec	1962	*His Majesty King Hussein of Jordan*
30 Jul	1963	Mr Julian E Ridsdale MP (Parliamentary Under-Secretary of State for Air)
17 Dec	1963	Air Chief Marshal Sir Edmund Hudleston GCB CBE ADC
28 Jul	1964	Air Chief Marshal Sir Walter G Cheshire KCB CBE ADC
17 Dec	1964	Air Chief Marshal Sir Charles Elworthy GCB CBE DSO MVO DFC AFC MA
24 Jul	1965	*Her Royal Highness The Princess Alexandra, The Hon Mrs Angus Ogilvy*
17 Dec	1965	Admiral Sir Varyl Begg GCB DSO DSC
18 Aug	1966	General Sir James Cassels GCB KBE DSO (Chief of the General Staff)
3 Mar	1967	Air Chief Marshal Sir Wallace Kyle GCB CBE DSO DFC ADC
26 May	1967	Air Chief Marshal Arjan Singh DFC (Chief of the Air Staff, Indian Air Force)
18 Aug	1967	General Lyman L Lemnitzer USAF (SACEUR)
1 Mar	1968	The Rt Hon Denis Healey MBE MP (Secretary of State for Defence)
18 Aug	1968	General A B Wolff (Chief of the Air Staff, Royal Netherlands Air Force)

28 Jan	1969	Air Chief Marshal Sir John Grandy GCB KBE DFC (Chief of the Air Staff)
1 Aug	1969	Generale Duilio S Fanali (Chief of the Air Staff, Italian Air Force)
27 Feb	1970	Air Chief Marshal Sir Donald Evans KBE CB DFC
31 Jul	1970	MRAF Sir Dermot Boyle GCB KCVO KBE AFC
26 Feb	1971	Air Marshal Abdul Rahim Kahn HQA SPK SBT (Chief of the Air Staff and Commander-in-Chief Pakistan Air Force)
25 Feb	1972	Admiral of the Fleet Sir Peter Hill-Norton GCB (CDS MOD)
16 Mar	1973	Air Chief Marshal Sir Denis Spotswood GCB CBE DSO DFC ADC (Chief of the Air Staff)

The passing out parade of No 101 Entry on 16 March 1973 was the last of what one might call the old-style flight cadet entry parades. From 1971 onwards the style of parade, review or inspection has undergone several changes. The following were the Reviewing Officers.

20 Aug	1971	Air Chief Marshal Sir Denis Spotswood GCB CBE DSO DFC ADC (No 1/70 GE)
15 Oct	1971	Air Chief Marshal Sir Andrew Humphrey KCB OBE DFC AFC (No 2 GE)
18 Feb	1972	MRAF Sir Charles Elworthy GCB CBE DSO MVO DFC AFC MA (No 3 GE)
21 Jul	1972	Air Chief Marshal Sir Frederick Rosier GCB CBE DSO ADC (No 4 GE)
15 Sep	1972	Air Chief Marshal Sir Christopher Foxley-Norris KCB DSO OBE MA (No 5 GE)
27 Oct	1972	Air Marshal Sir Leslie Mavor KCB AFC (No 6 GE)
19 Jan	1973	Air Marshal Sir John Lapsley KBE CB DFC AFC (No 7 GE)
22 Jun	1973	Air Chief Marshal Sir Neil Wheeler KCB CBE DSO DFC AFC (No 8 GE)
10 Aug	1973	Air Chief Marshal Sir Lewis Hodges KCB CBE DSO DFC (No 9 GE)
2 Nov	1973	Anthony Kershaw MC MP (No 10 GE)
11 Jan	1974	Air Marshal Sir Derek Hodgkinson KCB OBE DFC AFC (No 11 GE)
15 Mar	1974	Air Chief Marshal Sir Denis Smallwood KCB CBE DSO DFC (No 12 GE)
28 Jun	1974	*His Royal Highness The Duke of Kent* (No 13 GE)
16 Aug	1974	Air Chief Marshal Sir John Davis GCB OBE BA (No 14 GE)
1 Nov	1974	Air Chief Marshal Sir John Barraclough KCB OBE DFC AFC (No 15 GE)

17 Jan	1975	Air Chief Marshal Sir Anthony Heward KCB OBE DFC AFC (No 16 GE)
27 Mar	1975	HE General Hassan Toufanian (Vice Minister of War for the Imperial Iranian Government) (No 17 GE)
7 May	1976	*Prizes Parade* Air Marshal R D Roe CB AFC
19 Jul	1976	*The College Review, HRH Princess Alexandra The Hon Mrs Angus Ogilvy*
24 Jun	1977	*Prizes Parade* Air Chief Marshal Sir John Aiken KCB
23 Sep	1977	*The College Review, Her Royal Highness, The Princess Margaret, Countess of Snowdon*
22 Sep	1978	*The College Review,* Air Chief Marshal Sir Michael Beetham GCB CBE DFC AFC ADC (Chief of the Air Staff)
20 Sep	1979	Air Marshal Sir Keith Williamson KCB

Officers Commanding Electrical and Wireless School, Cranwell

No 1 Signals School, No 1 Radio School, No 6 Radio School.

1928-31	Group Captain A L Godman CMG DSO
1931	Group Captain R H Verney OBE
1934	Group Captain H Gordon-Dean AFC
1937	Group Captain J H Simpson
1940	Group Captain L T N Gould MC
1942	Group Captain C N Ellen DFC
1943	Group Captain G P Chamberlain OBE
1943	Group Captain S L Quine MC
1944	Group Captain G Thripp
1945	Group Captain N F Simpson
1946	Group Captain A T Monks
1948	Group Captain H A Evans-Evans
1950	Wing Commander H P Simpson
1951	Group Captain R J P Morris CBE
1951	Wing Commander A Foden
1952	Squadron Leader I E Hill

No 6 Radio School disbanded on 1 December 1952.

Training Aircraft RAF Cranwell[1]

Date in	*Date out*	*Type*	*Date in*	*Date out*	*Type*
Feb 1920	1927	Avro 504 K	Sep 1944	Jun 1954	Harvard
Feb 1920	May 1930	Bristol	1945	Apr 1949	Anson I
		Fighter	Jan 1949	May 1963	Anson
Feb 1920	May 1930	DH9a			19/21
Jul 1921	Sep 1927	Snipe	June 1948	Apr 1949	Vampire I
Feb 1922		Vickers	Jul 1948	Apr 1953	Prentice
		Vimy	Oct 1949	Sept 1954	Meteor T7
1927	Feb 1933	Avro 504 N	Jan 1952	1969	Chipmunk
Sep 1927	Nov 1933	Siskin	Jan 1954	Dec 1955	Balliol
May 1930	1934	Atlas	Sept 1954	Apr 1962	Vampire
May 1931	Mar 1933	Fairey Fox			T11
Feb 1933	1939	Tutor	Sep 1954	Feb 1961	Piston
Mar 1933	1939	Hart			Provost
Nov 1933	Feb 1936	Bulldog	May 1956	Mar 1966	Valetta
1934	1938	Audax	May 1957	Dec 1962	Varsity
Jan 1936	1939	Fury	Jan 1960	1961	Jet
May 1938	1942	Hector			Provost I
Jul 1938	1945	Oxford	Jul 1960	Nov 1979	JP3
Mar 1941	May 1945	Master	Dec 1961	Nov 1971	JP4
May 1943	Jan 1954	Tiger Moth	Jan 1966	Sep 1970	Varsity
Jul 1943	Apr 1945	Blenheim I	Jan 1971	Cont'g	JP 5/5A
Mar 1945	1945	Blenheim V	Dec 1973	Sep 1977	Dominie
Jan 1944	May 1945	Spitfire			

[1]From April 1976 to October 1977 the Flying Wing at Cranwell also housed the Jet Provost Mk 3A and 5A of the Central Flying School and the Vintage Pair (Meteor and Vampire).

Entries to RAF Cranwell Post-World War II

No	Enrolment Date	Passing Out Date
45	3 1 47	3 4 48
46	15 10 46 & 3 1 47	8 4 49
47	7 1 47	27 7 49
48	22 4 47	14 12 49
49	30 9 47	5 4 50
50	6 1 48	26 7 50
51	28 4 48	13 12 50
52	22 9 48	11 4 51
53	5 1 49	1 8 51
54	27 4 49	12 12 51
55	14 9 49	9 4 52
56	4 1 50	30 7 52
57	26 4 50	17 12 52
58	14 7 50	14 4 53
59	11 1 51	28 7 53
60	3 5 51	15 12 53
61	13 9 51	6 4 54
62	10 1 52	27 7 54
63	1 5 52	14 12 54
64	11 9 52	5 4 55
65	15 1 53	27 7 55
66	29 4 53	13 12 55
67	9 9 53	10 4 56
68	6 1 54	31 7 56
69	28 4 54	9 4 57
70	8 9 54	30 7 57
71	5 1 55	17 12 57
72	27 4 55	1 4 58
73	7 9 55	29 7 58
74	11 1 56	16 12 58
75	11 9 56	28 7 59
76	9 1 57	15 12 59
77	11 9 57	25 7 60
78	8 1 58	13 12 60
79	10 9 58	1 8 61
80	7 1 59	19 12 61
81	9 9 59	31 9 62
82	6 1 60	18 12 62
83	7 9 60	30 7 63
84	11 1 61	17 12 63
85	13 9 61	28 7 64
86	7 1 62	17 12 64
87	12 9 62	24 6 65
88	8 1 63	17 12 65
89	9 9 63	15 7 66
90	6 4 64	3 3 67
91	4 10 64	26 5 67
92	28 3 64	18 8 67
93	3 10 65	1 3 68
94	27 3 66	2 8 68
95	3 10 66	28 2 69
96	28 3 67	1 8 69
97	2 10 67	27 2 70
98	1 4 68	31 7 70
99	30 9 68	26 2 71
100	29 9 69	25 2 72
101	28 9 70	16 3 73

Entries to RAF Cranwell Post-1970

No	Entry Date	IOT Completion Date	No	Entry Date	IOT Completion Date
GE 1	1 9 70	20 11 70	22	28 7 75	31 10 75
2	27 10 70	29 1 71	23	20 10 75	30 1 76
3	22 2 71	28 5 71	24	12 1 76	15 4 76
4	9 8 71	21 10 71	25	29 3 76	9 7 76
5	13 9 71	31 12 71	26	2 8 76	5 11 76
6	1 11 71	11 2 72	27	25 11 76	25 2 77
7	17 1 72	28 4 72	28	14 3 77	24 6 77
8	28 2 72	9 6 72	29	25 7 77	28 10 77
9	29 8 72	1 12 72	30	7 11 77	12 2 78
10	30 10 72	9 2 73	31	—	—
11	8 1 73	13 4 73	32	24 7 78	27 10 78
12	12 3 73	22 6 73	33	30 8 78	15 12 78
13	14 5 73	17 8 73	IOT 34	4 12 78	16 3 79
14	28 8 73	30 11 73	IOT 35	12 2 79	18 5 79
15	5 11 73	15 2 74	IOT 36	9 4 79	13 7 79
16	14 1 74	26 4 74	IOT 37	21 5 79	21 9 79
17	25 3 74	5 7 74	IOT 38	18 6 79	19 10 79
18	28 7 74	1 11 74	IOT 39	13 8 79	14 12 79
19	22 9 74	3 1 75	IOT 40	22 10 79	29 2 80
20	17 11 74	28 2 75	IOT 41	19 11 79	28 3 80
21	20 1 75	2 5 75			

Equipment and Secretarial Wing RAF Digby

Cranwell Entry No	E & S Entry No	Enrolment Date	Passing out Date
49	No 1	15 9 47	26 7 50
50 & 51	No 2	29 4 48	13 12 50
52	No 3	22 9 48	11 4 51
53	No 4	5 1 49	1 8 51
54	No 5	27 4 49	12 12 51
55	No 6	14 9 49	9 4 52
56	No 7	4 1 50	30 7 52
57	No 8	26 4 50	17 12 52
58	No 9	12 9 50	14 4 53
59	No 10	10 1 51	28 7 53
60	E & S Wing[1]		

Officers Commanding

Group Captain R G Seymour CBE	1947-49
Group Captain D Lumgair	1949-52

[1] Integrated at Cranwell and passed out with 60 Entry 15 12 53

The RAF College Band

Old programmes and photographs confirm that there has been a military band associated with Cranwell since 1916. One photograph depicts the 'Royal Air Force' Band 1918, all in Royal Naval Air Service uniforms. The Band of the RAF College was commissioned on 1 April 1920[1] and principally supported college functions until 1973 when the administration and control of the band passed to the Ministry of Defence in a general centralization of RAF Bands. In the 60 or so years of its existence the band has undertaken many tasks of historical interest, and was the first band to broadcast from Cardiff in 1923. During this early period it is also reliably recorded that the band under the direction of Bandmaster A E Halford played a concert in an airship and, as far as is presently known, is the only established military band ever to do so. Travelling is part of a professional musician's lot; 'The day war broke out' the band was performing on Plymouth Hoe, and for the duration travelled the country entertaining various units. Affairs of State figure substantially in the diaries, the band being present at the Coronations of George VI and Queen Elizabeth II and also at the Jubilee celebrations of George V and Queen Elizabeth II. There is hardly a town that the band has not visited in its 60 years and recently it has been honoured with overseas visits to Canada, West Germany and Cyprus. The West Germany trip had some significance for the City of Lincoln which has a twinning with Neustadt and Weinstrasse, where on an evening off from duty the band played in the town centre to about 3000 citizens.

Since 1920 there have been seven Musical Directors among them being the legendary (George) A E Sims who was more famous as the Director of Music of the RAF Central Band. The present Director (1980) is Flight Lieutenant Robbie Wilkinson who was a playing member of the band in the 1960s.

[1] One of the musicians in that band was Trumpeter Bangay who became Squadron Leader W J Bangay MBE, Director of Music at Cranwell.

RAF Cadetships

Extracts from PAM (Air) 296, June 1979.

1 Because of the increasingly complex nature of defence systems and techniques the Air Force Board has decided that a high proportion of direct entrants to commissioned rank should be graduates. For this reason the RAF looks to the universities, polytechnics and other colleges for the young men and women who have the high personal and academic qualities required in an officer entering a full and progressive career with good opportunities of promotion to the higher ranks.

2 The GD Branch is the flying and executive branch of the RAF from whose senior ranks are drawn the Commanders-in-Chief and most members of the Air Force Board.

3 If you are selected for and accept a cadetship in the RAF you will be given a permanent commision and you will be required to serve in the RAF until you are 38 or have completed 16 years reckonable service from age 21 to qualify for retired pay, whichever is the later, ie to the 38/16 point. Your date of appointment to a commission as an acting pilot officer will be the date on which you first report to RAF College Cranwell or your UAS and sign your Joining Certificate.

4 RAF Cadetships are available in the General Duties Branch (Pilots and Navigators), General Duties (Ground) Branch (Aircraft Control and Fighter Control Specialisations), Photographic Interpretation Branch, Engineer Branch, Supply Branch, Administrative Branch (Secretarial, Education, Catering and Physical Education Specialisations), Security Branch (RAF Regiment and Provost Specialisations).

5 If you are a school leaver who has been provisionally selected for the award of a university cadetship, you will be required to attend a 14 day University Cadets Introductory Course at the Royal Air Force College Cranwell, during the summer vacation immediately preceding the start of your university course when the award of a cadetship has been confirmed. If you are already an undergraduate when awarded a university cadetship, you will be required to attend a similar course of 7 or 14 days duration during the next summer vacation following the award. Reports on your progress will be obtained at the end of each academic session.

After graduation University Cadet graduates attend an 18 week initial officer training course, normally at the Royal Air Force College Cranwell. On satisfactory completion of initial officer training graduates will progress to the RAF College or an RAF School appropriate to their branch for professional training.

Officers principally involved with the flying training of HRH The Prince of Wales

AOC-in-C Training Command Air Marshal Sir Leslie Mavor KCB AFC

Commandant RAF College AVM F D Hughes CBE DSO DFC AFC MA

Station Commander Group Captain G L Pendred

Chief Flying Instructor Wing Commander B Huxley

QFI to The Prince Squadron Leader R E Johns

OC HQ Squadron Squadron Leader J B Robinson AFC

Air Traffic Controller Flight Lieutenant M P Osborne

Lookout Aircrew Flight Lieutenant M A Gaynor

Civilian Batman Mr John Willows

Prizes and Awards RAF Cranwell[1]

The following annex based on a College Administrative Instruction lists the prizes and awards available to students who train at the Royal Air Force College and the Basic Flying Training School Cranwell. The use of the term 'cadet' embodies officer cadets and student officers. Reference to a Prizes and Awards Year (PAY) is to be interpreted as the period 1 January to 31 December. Annual prizes for the Department of Initial Officer Training (DIOT) will be presented at the College Reviews in 1980 and 1981. Annual awards for BFTS, DSGT and DAW will not be presented at the College Review. There will be a transitional period in three stages for the award of certain prizes in the DIOT until the last OCTU courses at Henlow have passed into the Service.

PRIZES AWARDED TO IOT STUDENTS

1 Course Prizes

Sword of Merit

This award will be a replica of a sword suitably mounted on a base bearing an engraved plate.

The sword will be awarded to the RAF student officer of each course who is approved by the D DIOT as having demonstrated outstanding ability, officer qualities and the potential to progress in the Service.

Sash of Merit

A Sash of Merit and a silver and engraved brooch were originally awarded to the WRAF Graduate Entrants in 1971 as an annual award. For policy reasons this was discontinued after 1972 and only two WRAF officers possess brooches. The cost of the dies was provided by Air Cdre P F Marshall CB OBE Director of the Women's Royal Air Force from the DWRAF's Miscellaneous Fund and Training Command, now Support Command, provided the money for the purchase of a sash and a badge. Events overtook the actual putting of this money into use and only a sash was obtained. The Sash of Merit has now become a course award with a brooch as a prize.
The Sash of Merit will be awarded to the WRAF Cadet of each course who is approved by the D DIOT as having demonstrated outstanding ability, officer qualities and potential to progress in the Service.

[1] As at June 1980

The Hennessy Trophy and Philip Sassoon Memorial Prize

This trophy was given to the College in 1960 by Flight Lieutenant Dennis Hennessy who at that time was a Flight Commander in the Flight Cadet Training Wing, the trophy being known as the Hennessy Leadership Training Trophy for inter-squadron competition. When the training of flight cadets ceased the award fell into abeyance until 1975 when, with the approval of the donor it became an individual award for student officers and continued to remain so being linked with the Philip Sassoon Memorial Prize from 1 January 1980. The donor retired from the Service in 1976 in the rank of squadron leader.

The Philip Sassoon Memorial Prize was given by the late Sir Philip Sassoon former Under-Secretary of State for Air and was first awarded to the flight cadets in 1948 and the Graduate Entrant officers in 1971. The wording of the terms of the trust state that the prize must be of intrinsic value capable of being engraved or having an engraved plate attached and that the engraving must include the College crest. In addition the winner is permitted to add to the prize money up to twenty-five per cent of its value from his/her own sources.

The trophy and prize will be awarded to the RAF or WRAF cadet who is approved by the D DIOT as the best all-round cadet of each course.

The cadet eligible for this award is either the Sword of Merit or Sash of Merit winner.

The British Aircraft Corporation Trophy

This trophy which is a silver-plated replica of a Lightning aircraft mounted on a half-globe was first presented to the OCTU Henlow in 1969 by the British Aircraft Corporation together with an endowed fund to provide individual commemorative plaques for the winners. Over the period of time the endowment fund has expired and the cost of the plaques is now borne by the OCTU Endowment Fund. At the time the trophy was given Marshal of the Royal Air Force Sir Dermot Boyle GCB KCVO KBE AFC RAF was Chairman of the British Aircraft Corporation.

To be awarded to the cadet, whether RAF, WRAF or foreign and Commonwealth with the highest mark for professional studies on each course.

A plaque bearing an aircraft in bas relief is retained by the winner.

The Overseas Students' Prize

This award consisting of a plaque bearing the College arms will be awarded to the Foreign and Commonwealth (F and C) cadet on each course who has the best overall performance in leadership, officer qualities and professional studies.

The Alasdair Black Memorial Trophy and Prize

The trophy and prize were originally awarded to the Royal Air Force Technical College in 1955 by Lady Tedder in memory of her son Flight Cadet Sergeant Alasdair S H Black who was killed in a flying accident in 1950 whilst in training at the Royal Air Force College. The trophy and prize were transferred to the RAF College on the merger of the College in 1966 and awarded to the flight cadet who achieved the highest standard in all aspects of General Service Training. On the cessation of flight cadet training in 1973 the trophy was handed over to the custody of HQ TC and re-allocated to the OCTU at Henlow as an annual award to a cadet on similar terms to those whilst in use at Cranwell. The original prize was a copy of the book *'With Predjudice'* by the late Lord Tedder which was suitably bound and inscribed, but this book is no longer obtainable. A replacement book is being sought by the D DIOT.

The trophy and prize is awarded to the RAF or WRAF student officer of each course who submits the best essay on War Studies.

2 Annual Prizes

The Queen's Medal

The medal was first awarded to flight cadets in 1952 being a continuation of the King's Medal first awarded in 1933. In 1974 Her Majesty the Queen graciously approved the continuity of the award in favour of the Graduate Entrant students. The cost of the minting and embossing of the medal is borne by the Privy Purse. The medal bears only the year of the award, and the name and initials of the winner are embossed around the edge.

The Queen's Medal is awarded to the RAF or WRAF cadet in each PAY who in the opinion of the AOC and Commandant proves themselves to be the most outstanding cadet of the year.

The Sword of Honour

The Sword of Honour is awarded by the Air Force Board of the Defence Council and was first awarded to the flight cadets in 1921 as an entry award. The Graduate Entry students were awarded the sword as a course award in 1971 to 1975. Escalating costs necessitated that the award became an annual one and from its introduction in 1921 to 1979 it was awarded only to students who had completed both their IOT and IST at the College.

The Sword of Honour is now awarded to the RAF cadet in each PAY who in the opinion of the AOC and Commandant proves himself to have most distinguished himself in leadership and in general influence for the good of the College during his IOT at Cranwell.

Sash of Honour

This award is being introduced with effect from 1980 in order that recognition can be given to the WRAF cadets at the Royal Air Force College equivalent to that received by the Sword of Honour winner. A sash was previously awarded annually to WRAF Graduate Entrants in 1971 known as the Sash of Merit but due to a change in training policy ceased in 1972 and there are only two WRAF officers who received the original award and the silver brooches that were retained by the winners. The money for the purchase of the dies for the production of the brooch was provided by Air Commodore P F Marshall CB OBE Director of the Women's Royal Air Force from the DWRAF's Miscellaneous Fund. The brooch consisted of a pair of wings centred with an enamelled College crest surmounted by a crown.

The Sash of Honour and Brooch is now awarded to the WRAF cadet in each PAY who in the opinion of the AOC and Comdt proves herself to have most distinguished herself in leadership and in general influence for the good of the College during her IOT at Cranwell.

The R S May Memorial Prize

Flight Lieutenant Richard Steven May, a former flight cadet who graduated from the College in December 1950 was killed in a flying accident in February 1958. In his Will he bequeathed a sum of money to create a Trust Fund to provide a sum of money annually to be awarded to the winner of the Sword of Honour, at that time an entry award. In 1973 the award was continued in favour of the Graduate Entrant students until 1975 when the Sword of Honour became an annual award. The annual income from the trust fund was considered too great to award it to one individual per year and it was therefore shared between the Queen's Medal and Sword of Honour winners.

The value of the prize is determined at the commencement of a PAY by the annual income which is paid into the Royal Air Force College Mess account twice yearly by the United Services Trustee and it is further determined by sharing the annual income equally between the recipients.

The persons eligible for a share of this prize are the winners of the Queen's Medal, Sword of Honour and Sash of Honour.

The Prince Bandar Trophy and Prize for War Studies

A sum of money and a silver trophy have been donated to the Royal Air Force College by HRH Prince Bandar bin Sultan bin Abdul Aziz Alsaud of Saudi Arabia to provide an Annual Prize for War Studies.

Prince Bandar trained at Cranwell from 28 March 1967 to 1 August 1969 as a flight cadet in 'D' Squadron of No 96 Entry.

The actual terms of this award will be published as soon as they have been decided.

The Argentine Air Force Senior Prize

Whilst this award is not truly a College prize it is only available to students from the Royal Air Force College.

The Argentine Air Force pursuant to a decree issued by the Argentine Government awarded an annual prize to the RAF flight cadet, irrespective of branch, who attained certain high standards in all aspects of his training at Cranwell, and the prize has always been awarded to a Queen's Medal winner. The prize is a replica of the Sword used by the South American Liberator General San Martine, the leader during the wars of independence in Argentine and Peru and the liberator of Chile from the Spanish control. General San Martine died in 1850. The prize was first awarded to the flight cadets in 1956, and when flight cadet training ceased in 1973 the Argentine Government agreed to the continuation of the award to the Graduate Entrant students on the same terms, the first award being made in 1974.

The award will continue to be made to the Queen's Medal winner if an RAF cadet who is automatically nominated by MOD.

There is also a Junior Prize awarded to apprentices. Both the Senior and Junior awards are presented by the Argentine Ambassador or his representative at an RAF station.

PRIZES AWARDED TO STUDENTS ON BFTS CRANWELL

3 Course Prizes

The Hicks Memorial Trophy

This trophy was presented to the College in memory of Flight Cadet John Lawrence Orly Hicks by his parents. Flight Cadet Hicks was killed in a motor accident in 1939. Originally awarded as a prize for aerobatics this silver cup was awarded for the first time in 1954 and then went out of use after 1955. In 1958 it was once again awarded to the flight cadet who was the runner-up to the R M Groves Memorial Prize Winner. In 1963 the terms of the award were again changed and it became the Flying Wing Ground School Prize and in 1971 under the same terms it was awarded to the Graduate Entrant students.

The cup is now awarded to the RAF pilot who is placed first in order of merit in the Basic Flying Training School Ground School.

The Dickson Trophy

The Tours Schneider Cup was won at Tours in 1910 by Captain Bertram Dickson of the Royal Horse Artillery later to become a member of the Royal Flying Corps. On his death the cup was presented to the College by his sister Mrs Wills Gordon, Chevalier de la Legion D'Honneur, as an award for a degree in excellence in connection with flying and was first awarded to the flight cadets in 1948 for proficiency in instrument flying. In 1953 this award was linked with the Michael Hill Memorial Prize and awarded for proficiency in applied flying. In 1971 the Dickson Trophy was separated from the Michael Hill Memorial Prize when it was also awarded to the Graduate Entrant students until the end of the flight cadet training in 1973 when the awards were once again linked. However, in 1978 the awards were again separated for financial reasons.

The Dickson Trophy is now awarded to the RAF pilot in the Basic Flying Training School, who, on the completion of his training at Cranwell, has shown most proficiency in applied flying.

The Battle of Britain Trophy

At the instigation of Air Commodore A C Deere DSO OBE DFC in 1963 when he was Assistant Commandant of the College, the Battle of Britain Fighter Pilots Association agreed to provide a trophy for skill in aerobatics to perpetuate the Battle of Britain in association with the College where many of the pilots were trained. The trophy was first awarded to the flight cadets in July 1963 and to the graduate entrant students in 1971.

This trophy is awarded to the RAF, Navy or Army pilot who on completion of his training in BFTS at Cranwell is adjudged to be the best aerobatic pilot on his course. This trophy is the only College award open to all three Services by virtue of the fact that pilots of all three Services fought side by side in the Battle of Britain.

The R M Groves Memorial Prize

This prize was donated by the Groves family, whose interest in the College is still closely maintained to this day, in memory of Air Commodore Robert Marsland Groves, CB DSO AFC who was killed in a flying accident in Egypt in 1920. Air Cdre Groves was the first officer to hold the appointment of Vice Chief of Air Staff. The principal donors of this prize were the mother of Air Cdre Groves and his uncle Mr Gribble Groves. Other members of the family actively connected with the award were Major Keith Groves OBE, brother of Air Cdre Groves and Gp Capt H M Groves, an ex-flight cadet who graduated in 1923, son of the Air Cdre. All these persons are now deceased and the family is now represented by Mr N E Abbott, grandson of Gp Capt Groves and Mrs I Wight sister of Gp Capt Groves, but money has always been bequeathed to maintain the value of the award. The terms of the trust which is administered by MOD (Air) clearly states that the award must be made to an RAF student and that the book presented shall contain a photograph of the Air Commodore and a facsimile of his signature, and any balance of money after the purchase of the book shall be paid to the winner. The award was originally made to the flight cadet who, on passing out of the College was the best all-round pilot of his term. The first award was made in 1921, and to the Graduate Entrant students in 1973. From 1929-1977 the award was linked with the Kinkead Trophy.

The prize is awarded to the RAF pilot who is placed first in the combined order of merit for flying and associated ground school subjects, at the end of his training at the BFTS at Cranwell.

4 Annual Prizes

The Kinkead Trophy

The trophy was donated to the College by the family and friends of the late Flight Lieutenant S M Kinkead DSO DSC DFC who was an instructor at the College from 1920-24. Flight Lieutenant Kinkead was killed in 1928 whilst attempting to break the world air speed record over the Solent in a Supermarine S5 aircraft the forerunner of the famous Spitfire. A model of the S5 crowns the magnificent trophy. The trophy was first awarded in 1929 and until 1958 was presented to the Flight Cadet Squadron who produced the winner of the R M Groves Memorial Prize. In 1958 it was decided that to give greater recognition to the award it was linked with R M Groves Memorial Prize as an individual award thus permanently linking the names of two outstanding RAF officers. The award was continued into the Graduate Entrant era, the first award to a Graduate Entrant student in 1973 and remained a course trophy until 1976 when it was decided that some recognition should be made of the best R M Groves Memorial Prize winner of the year and it was first awarded as an annual trophy in 1977.

The trophy is awarded to the RAF pilot, trained at the BFTS Cranwell who proves himself to be the best of the R M Groves Memorial Prizes in a PAY subject to the AOC and Commandant's approval.

The Michael Hill Memorial Prize

In 1951 Mr R H Hill offered to provide a trust to finance an annual prize to be known as the Michael Hill Memorial Prize in memory of his son Squadron Leader Michael Hill who was killed in World War II whilst serving with No 19 Squadron. In 1952 the first award was made, the prize being linked with the Dickson Trophy for proficiency in instrument flying. In 1962 the terms of the award were amended to proficiency in applied flying and these terms have been retained until the present day. The award was first made to a Graduate Entrant student in 1973. Originally the purchasing power of the money enabled the recipient to buy an article which could be engraved but in 1974 it was necessary to make the prize a book until finally in 1978 it was decided that it should become an annual award with effect from 1979 in order that the winner could have something worthwhile to show for his achievement, the annual income being sufficient to obtain a small engraved article.

The prize is awarded to the RAF pilot trained at the BFTS Cranwell who proves himself to be the best Dickson Trophy winner of the PAY subject to the AOC and Commandant's approval.

DEPARTMENT OF SPECIALIST GROUND TRAINING

5 Course Prizes

The Whittle Prize

This prize was founded by Air Commodore Sir Frank Whittle KBE CB MA CEng FRS, in 1954 to be awarded each year to the cadet at the Royal Air Force Technical

College, Henlow, who submitted the best essay on a technical subject. On the absorption of the Technical College into the Royal Air Force College in 1966 the award continued but the terms of the award were changed to being awarded to "the Honours BSc degree student who submitted the best thesis in conjunction with his project". With the demise of the flight cadet scheme a further change occurred. Sir Frank Whittle increased the value of the award from £20 to £30 and authorised MOD (Air) and the AOC and Commandant to set the terms of the award without referring their proposals back to him. From 1975 the prize was awarded to the Aerosystems Engineering Course student who submitted the best individual project thesis and very shortly afterwards the terms were again changed to those in use today.

This prize is awarded to the student on the Aerosystems Engineering Course who is adjudged to be the best student in terms of progress in studies and contribution to the success of the course as a whole.

The Royal United Services Institution Prize

This prize came into being in 1947 when the Council of the Royal United Services Institution on the graduation of each flight cadet course from the College presented free and full membership of the Institution to the flight cadet whose work on his selected subject was considered by the Commandant on the advice of the Director of Studies to be the most meritorious. The award was first made in 1948 and at that time there were three entries graduating each year. In 1956 when graduations fell to two per year the RAF Technical College at Henlow requested that the third award be allocated to them and the RUSI Council agreed. In 1962 the terms of the award were revised to "awarded to the Flight Cadet in the 'A' Academic Stream who showed the greatest all round attainment in all subjects based on the result of the final examinations". A further change was made in 1967 when the award was given to the runner up in the Whittle Prize competition and has remained so to date although the terms of the Whittle Prize have been revised.

The RUSI prize is awarded to the student who is adjudged runner up in the Whittle Prize competition.

The Halahan Trophy and Prize

In 1969 Miss Patricia Halahan, daughter of the late Air Commodore F C Halahan CMG OBE DSO MVO, Commandant of the Royal Air Force College from 1926-29, who retired in the rank of air vice-marshal in 1943 and died in 1965, offered a silver cup to the College to be awarded as a prize. The cup was one given to her father on his retirement. In addition, Miss Halahan gave monetary awards for purchase of books up to the value of £15 and £10 for the winner and the runner up respectively. The first awards were made in 1971.

The trophy and first prize are awarded to the engineer officer, irrespective of specialization, who achieves the best all-round performance on each Maintenance Engineering Course, and the runner up is the officer who achieves the best all-round performance in the specialization alternative to that of the winner.

The Royal New Zealand Air Force Trophies and Prizes

There are two RNZAF trophies which are identical. The first trophy was presented to the College in 1957 as a flight cadet award. The second trophy was presented to the RAF Technical College, Henlow in 1961 where it was awarded to the cadet of the final year who showed the greatest originality in conception, planning, leadership and determination in carrying out a vacation activity of an adventurous nature. The College prize was for Commonwealth and War Studies originally and later for proficiency in Military Studies and Humanities subjects. Both the Cranwell and Henlow awards were accompanied by book prizes financed by the Royal New Zealand High Commission. When the Technical College was absorbed into the RAF College in 1966 the Henlow trophy continued to be used by the Department of Engineering. When the training of flight cadets ceased at Cranwell the College Trophy and prize were handed over for use by the Department of Engineering, the first dual awards being made in 1974 to students of each specialization on the Maintenance Engineering Courses as they still are to date.

The trophies and prizes are awarded to a student of each specialization on a Maintenance Engineering Course submitting the best paper on a selected engineering subject.

The Beckwith Trophy

This trophy is a silver rose bowl given to the College by Mrs H Beckwith in memory of her husband Air Vice-Marshal W F Beckwith CBE BSc Eng ACGI CEng FIMechE FRAeS who died in 1971. From 1954-57 he was Assistant Commandant at the Royal Air Force Technical College at Henlow.

The trophy is awarded to a student of each Initial Engineering Course who, irrespective of specialization, submits the best paper on a selected engineering subject.

The Supply Cup and The Secretarial Cup

To date the history and origin of the cup used for Supply studies has not been traced but it is known that the Secretarial Cup was originally presented to the College by the President of the British Motor Corporation and known as the Harriman Cup to be awarded to the winner of the Flight Cadet Road Safety Competitions. In 1971 when Graduate Entrant students started to train in Supply and Secretarial duties at Cranwell there were no awards available as the MOD awards were still being used by the flight cadets, therefore these cups were awarded and known as the Prize for Supply/Secretarial Studies. With the departure of the last flight cadets in 1973 MOD withdrew their prizes on the grounds that there were insufficient number of GE students being trained at Cranwell. The titles of the awards were changed in 1975 to those we use today and all students whether GE or DE were eligible.

The Cups are awarded to the student of each specialization who achieves the highest standard in their professional studies.

The Chicksands Cup and Prize

This trophy was originally presented to the Royal Air Force Technical College Henlow by the United States of America Air Force Base at Chicksands round about 1955. The first award was made to the senior technical cadet of No 1 Entry who had most distinguished himself in academic and professional studies. There is no further knowledge of this trophy until the Technical College was absorbed by the RAF College to become the Department of Engineering in 1966. From 1967 onwards it was awarded annually to the flight cadet taking the BSc Honours Sandwich Course in Electrical/ Mechanical Engineering who achieved first place in the order of merit. The winner also received a book prize to the value of £5. In 1970 the word 'Sandwich' was deleted from the citation. The sum of £5 was received from an HMSO vote. In 1973 with the graduation of the last flight cadet entry the award became available to the Graduate Entrant student the Cup being awarded to the Graduate Entrant student of each IEC who submitted the best paper on an engineering subject, the prize being split into 2 parts £10 for each student who submitted the best paper of the intake year on an electrical engineering subject and a mechanical engineering subject. In 1975 the Cup and the Prize were re-united to be awarded annually to the Graduate Entrant student who achieved the highest standard in professional studies on the Initial Engineering Courses, irrespective of speciality, in the intake year and in 1978 the award was thrown open to all students on all IECs.

The Chicksands Cup and Prize is awarded to the student who achieves the highest standard in professional studies irrespective of specialization on the Initial Engineering Courses in each PAY.

The Beckwith Prize

This prize is awarded annually by Mrs H Beckwith in addition to the course trophy in memory of her husband the late Air Vice-Marshal W F Beckwith CBE BSc Eng ACGI CEng FIMechE FRAeS.

The prize is awarded to the student who in a PAY has submitted the best essay on a selected engineering subject, having been awarded the Beckwith trophy for that essay.

The RAF College Prize for Supply Studies and The RAF College Prize for Secretarial Studies.

When flight cadet training ceased in 1973 the flight cadet awards provided for by MOD (Air) were lost on the grounds that insufficient numbers of students were trained in these professions at Cranwell. To offset this loss of prizes the AOC and Commandant agreed to the above prizes being introduced and awarded to the Graduate Entrant student who was considered to have achieved the highest standard in Supply/Secretarial Studies in the intake year, provided that in the opinion of the AOC and Commandant the results were of a sufficiently high standard. The first award was made in 1976 and in 1978 the prizes were made available to all students on the SOCs and ISCs.

DEPARTMENT OF AIR WARFARE PRIZES

7 Course Prizes

The Andrew Humphrey Memorial Gold Medal

The gold medal is provided by Lady Humphrey in memory of her husband the late Marshal of the Royal Air Force Sir Andrew Humphrey GCB OBE DFC AFC RAF who at the time of his death was Chief of the Defence Staff and was first awarded in December 1978.

The medal is awarded to the best overall student on the GD Aerosystems Course suitably inscribed with the winner's name.

The Aries Trophy

The Specialist Navigation Association was formed by the College of Air Warfare at RAF Manby in 1965 and awarded the Specialist Navigation Association Trophy to the student on the Specialist Navigation Course who submitted the best personal project. In 1969 the title of the course was changed to the GD Aerosystems Course and in 1973 the title of the Association was changed to the Aries Association and the trophy became the Aries Trophy.

The trophy is awarded to the student on each GD Aerosystems Course who produces the best personal project. The trophy is retained by the College and the winner receives a small replica which is suitably engraved.

PRIZES FOR WHICH ALL STUDENTS IN DIOT, BFTS AND DSGT ARE ELIGIBLE

Lowe-Holmes Trophy

In 1972 Group Captain D H Lowe-Holmes RAF donated £1,000 to the College for the purchase of a trophy to the value of £250 the remaining money to be invested as a trust fund to provide income for biannual grants to the winners of the trophy. The year is divided for sporting purposes into a summer season and winter season. The full title of the award is 'The Group Captain D W Lowe-Holmes Award for Sport' and was first awarded at the end of the 1973-74 winter season.

The award is made to the student officer or officer cadet who, during his/her service at the RAF College, is considered to have given by his/her active, cheerful and sportsmanlike participation, the greatest service to sports and sportsmanship at the College. These conditions offer equal opportunity to sportsmen and sportswomen of all levels of ability or participation.

Glossary of Code Names and Abbreviations

ACAS	Assistant Chief of the Air Staff
AC2	Aircraftman Second Class
AFB	Air Force Board, eg, AFB(66)21 is the reference of an AFB meeting (paper) in 1966
AFS	Advanced Flying School
ANS	Advanced Navigation School
AMP	Air Member for Personnel
AMSO	Air Member for Supply and Organization
AOC	Air Officer Commanding
BFTS	Basic Flying Training School
CAS	Chief of the Air Staff
CFS	Central Flying School
CH	Chain Home (radar)
CHL	Chain Home Low (radar)
CPO	Chief Petty Officer
CNAA	Council for National Academic Awards
CTE	Central Training Establishment
'Cranwell Express'	The passenger and goods train which ran between Cranwell and Sleaford
DE	Direct Entry
DFT	Director of Flying Training
DGO	Director General of Organization
DGT	Director-General Training
Dip Tech	Diploma in Technology
DPS	Director of Personal Services
DTT	Director of Technical Training
DWB	Director of Works and Buildings
EVT	Educational and Vocational Training
FTS	Flying Training School
GD	General Duties
GDE	Graduate Direct Entrant
GE	Graduate Entry
HQ	Headquarters
HND	Higher National Diploma
IOT	Initial Officer Training

'King Rock'	Leadership and endurance training exercises frequently in Germany
LAA	Leading Aircraft Apprentice
LAC	Leading Aircraftman
MT	Mechanical Transport
OATS	Officers' Advanced Training School
OC	Officer Commanding
OTU	Operational Training Unit
'Piloteer'	Magazine of the Central Training Establishment RNAS Cranwell
PUS	Permanent Under Secretary
RFC	Royal Flying Corps
RNAS	Royal Naval Air Service
R/T	Radio Telephony
SFTS	Service Flying Training School
SGE	Single Gate Entry
SS	Submarine Scout (airship)
'Superna Petimus'	Roughly translated means 'We seek things that are above'
UAS	University Air Squadron
UCG	University Cadet Graduate
WAAF	Women's Auxiliary Air Force
WOp	Wireless Operator
WOp Mech	Wireless Operator Mechanic
WRNS	Women's Royal Naval Service
W/T	Wireless Telegraphy

Sources of Information

The Operations Record Books (F540) of the RAF College and other units at Cranwell provided the bulk of the information used in this book. The principal ORBs with their Public Record Office classification are given below:

Air 28/175	RAF Station Cranwell 1936—1945 (including RAF Caister 1943/44)
Air 28/1011	RAF Station Cranwell 1946—1950
Air 29/699	RAF College Cranwell (including RAF College SFTS and No 17 SFTS) 1920—1949
Air 29/1145	No 19 Flying Training School May 1945—Apr 1947
Air 29/630	No 2 Central Flying School
Air 29/705	No 3 (C) OTU Oct 1940—Dec 1943
	Electrical and Wireless School (later No 1 E and W School, No 1 Signals School No 1 Radio School 1915—1945
Air 29/1853	No 6 Radio School Oct-Dec 1950
Air 29/725	No 1 Radio School (later No 1 RDF School No 8 Radio School 1941—1946
Air 25/510	HQ No 21 Group Dec 1938—Jun 1945
Air 29/1157	OATS (later No 1 OATS) Mar 1944—Dec 1945
Air 29/602	Beam Approach School Jan—May 1944
Air 29/727	School of Clerks Accounting
Air 29/711	School of Store Accounting and Storekeeping 1929—1944
Air 29/833	Supplies Depot 1935—1950
Air 29/1831	No 3 ITW (later No 3 ITS) Jan 1951—Mar 1953

Papers concering the following units were studied. No F540s as such exist for them:

Aircrew Transit Centre	Jan 1952—July1953
RAF Long Range Flight	1930—1938
RAF Selection Board	May—July 1953 July 1962—Mar 1964

RAF Cadet College Magazine 1920—1980 (later became RAF College Journal)

"The Piloteer"—Magazine of RNAS Station Cranwell 1915—1919

Boys' Wing Magazine 1922—1926

Diary of HM King George V and Letters of Prince Albert to The King and The Queen 1918.

Private papers of Lord Trenchard at Abdale House.

Minutes of the Old Cranwellian Association from 1926.

Newspaper cuttings files in College Hall Library Cranwell.

Log books of J H Lidderdale (in possession of Mrs Pam Smith).

Fifty Years of Cranwell – a pamphlet – covering 1920 – 1970.

The Old Hangar Church – a pamphlet – 1960.

A Short History of The Lodge, 1967.

List of Graduates Feb 1920 – Dec 1962.

Files and papers in Air 1, Air 2 and Air 8 classifications in the Public Record Office.

Papers not yet released to the Public Record Office under the Public Information Act of 1964 are preserved mainly in the Air Historical Branch (RAF). The principal sources of information were Air Council (later Air Force Board) Minutes and Conclusions and associated papers. The folders of the Chief of the Air Staff and Vice-Chief of the Air Staff were also studied. References are made throughout the text to various important committees and their reports and should be studied in their context in the narrative.

Certain published books provided helpful background information:

The Parish of Cranwell, Harold Ison, Sleaford.

A History of the RAF College Cranwell, R de la Bere, Gale & Polden, 1934.

King George VI: His Life and Reign, J W Wheeler-Bennet, Macmillan, 1958.

Lincolnshire, White, 1892.

Ships of the Royal Navy, J J Colledge, David & Charles, 1969.

Solitary in the Ranks, H Montgomery Hyde, Constable, 1977.

Lincolnshire, Pevsner and Harris, Penguin Books, 1964.

British Air Policy between the Wars, 1918 – 1939, H Montgomery Hyde, Heinemann, 1976.

Trenchard: Man of Vision, A Boyle, Collins, 1962.

The World Crisis: The Aftermath, W S Churchill, 1929.

All the Days of My Life, S P B Mais, Hutchinson, 1937.

Notes

1 Naval beginnings–The Royal Naval Air Service 1915–1918

1 *A History of the RAF College Cranwell*—R de la Bere, Gale and Polden, 1934.

2 *The Parish of Cranwell*—Harold Ison, published by F M Morton & Son Ltd., Sleaford.

3 File numbered Air 2/19 in Public Record Office.

4 Letter from Lt Col F H Sykes to MT4 at War Office on 15 August 1913.

5 F C Halahan was a naval officer who became an expert in AA gunnery. He took part in organizing the AA Corps in London in 1914. After serving in the Air Department with the RNAS followed by war service at Dunkirk, he transferred to the RAF in 1918. He became Commandant at Cranwell from 1926 to 1929. His daughter, Pat Halahan, was a teen-aged young woman at Cranwell and gave the author some lively accounts of life there in the late 1920s.

6 Daedalus. HMS *Thunderbolt* built at Poplar and launched in 1856 was never completed for sea but made a Floating Battery. In 1873 she was attached to a temporary pier at Chatham for use of ships fitting out. On 16 December 1915 she was re-named Daedalus and became the Flying School Depot Ship (RNAS) in the Medway. In January 1920 she reverted to a hulk and was used for berthing tugs. She was rammed and sunk by a tug in 1948 according to J J Colledge *Ships of the Royal Navy*, David & Charles, 1969.

7 No evidence of badge or sign could be found by the author to indicate that Cranwell itself was known as Daedalus. The Commandant's office still has an ashtray inscribed 'RNAS Central Training Establishment'.

8 File DW 24626/15, Public Record Office Air 2/19.

9 The four sources are:

a *Cranwell—a Retrospect 1916—1932*, by J W Robinson, Postmaster, published in RAF College Journal, 1932.

b Reminiscences by Wing Commander A E F McCreary.

c Notes on Cranwell by Mr S Godson, formerly groom to the Commandant.

d *Cranwell in the Old Days* by Chief Petty Officer W H Whitlock, published in RAF College Journal 1936.

To these sources can be added an article, *The Birth of Cranwell*, by Avion, printed in *the Piloteer*, June 1917.

10 Article published in RAF College Journal,

11 Information on J H Lidderdale supplied mainly by his widow, Mrs Jill Lidderdale and by her daughter, Mrs M Smith (Air Staff Briefing branch MOD). Lidderdale was almost a legend in himself and deserves the fuller treatment given to him in Annex II.

12 Files in PRO under Air 2/19 and Air 2/28. Also letter to author from Captain Sir Anthony Thorold in 1979.

13 There is an excellent account of the Cranwell-Sleaford railway in the RAF College Journal of November 1953.

14 Captain Sir Anthony Thorold RN (Retd) the present head of the family confirmed in a letter to the author, dated 8 February 1979, that the total land sold by his grandfather amounted to 1416 acres, 1 rod, 12 poles and the sale document was signed on behalf of the Air Ministry by Mr Winston Churchill.

15 File Northern 8/2960 in Air 2/28 in PRO.

16 In this connection see also 'Statement of Aeroplanes and Engines handed over to RNAS Training Establishment on transfer of training of Naval Pilots from Central Flying School' in Air 2/13 in PRO.

17 Interview with Mr A H Mather JP of Boston. The pilot, according to Mr Mather, was Flight Lieutenant Ray Hinchliffe who was later lost over the Atlantic.

18 RA GV.

19 The Hotel Cecil, London, was the first headquarters building of the Air Board. It stood near where the present Shell Mex House stands.

2 Connections with the Royal Family

1 RA GV diary.

2 *King George VI. His Life and Reign* by J W Wheeler-Bennet, p. 103.

3 RA GV CC10/268.

4 RA GV AA60/280.

5 RA GV CC10/274.

6 RA GV AA60/280.

7 This was R E C Peirse, later Air Chief Marshal Sir Richard Peirse.

8 RA GV CC10/271.

9 RA GV Diary.

10 RA GV AA60/301.

11 RA GV AA60/302.

12 op cit Wheeler-Bennett, p. 112.

13 RA GV CC10/283.

3 Early Years at Cranwell

1 Churchill, *The World Crisis: The Aftermath* (1929), p. 52.

2 H Montgomery Hyde, *British Air Policy between the Wars, 1918–1939*, p. 52.

3 Cmd 467/1919.

4 Air 8/79.

5 From Lord Trenchard's private papers at Abdale House.

6 Air 8/10.

7 Extract from file A.13339 in Air 2/100. The other members of the Committee under Lord Hugh Cecil were:
Captain W Wedgwood Benn, DSO, DFC
S P B Mais, Esq
Air Cdre J G Hearson, CB, DSO (Deputy Director of Training)
Group Captain A M Longmore, DSO
E Halford Ross, MD.

8 Minute by CAS on file A.19273 in Air 2/100.

9 Letter from Ludlow-Hewitt to General Scarlett at Cranwell on 3 September 1919.

10 From Trenchard's private papers at Abdale House. Longcroft soon found himself in trouble when at a later date he proposed hunting and Trenchard was bent on business only!

11 For all references to entries the List of Graduates, 1920–1962 is indispensable.

12 Vol 1 No 1. RAF Cadet College Magazine, September 1920.

13 Air 2/9381. Candidates for a King's Cadetship had to pass the normal medical examination and the Civil Service Commission Entrance Examination. Eligible candidates were the sons of officers who had fallen in action, died of wounds received in action or of diseases contracted abroad or killed as a result of a flying accident whilst on duty.

14 Sir Charles Wakefield (1859–1941) 1st Viscount Wakefield of Hythe. *Who Was Who* lists his benefactions.

15 The chemical treatment of the fabric for aircraft wings and fuselage to increase tautness and efficiency took place in the 'dope shed'.

16 S P B Mais, *All the Days of My Life* (1937), p. 83-117.

17 Group Captain F E Nuttall, CBE, was the distinguished and respected College Mess Secretary 1953—1965. He lived in Cranwell Village in his retirement and maintained close links with the College.

18 For this and other reminiscences the author is indebted to Group Captain A H W J Cocks (1923 Entry).

19 From Lord Trenchard's private papers at Abdale house.

20 Letter on file 322570/20/S.7 dated 22 September 1920.

21 Letter 190511/20/S.6 dated 23 September 1920.

22 S P B Mais, *The Happiest Days of My Life* (1953).

4 RAF Cadet College Life 1920–1929

1 List of Graduates—Old Cranwellian Association Publication—5 February 1920 to 18 December 1962.

2 CAS Minute to DTO on file A49482.

3 Box No 140 in AIR 2.

4 Letter dated 23 February 1923 in Lord Trenchard's private papers.

5 Letter from T B Marson to Longcroft on 27 May 1922. 'The CAS is having sent down to you the banners of Nos 2, 3, 4, 5 and 6 Squadrons. Lady Leighton and Lady Henderson caused them to be made.'

6 The Old Hangar Church, 1960. The Church of St Michael and All Angels, 1968. Booklets produced by incumbents at Cranwell.

7 The first committee consisted of the following former cadets:

Plt Off G R Beamish	(23B)
Fg Off J G Hawtrey	(20A)
Fg Off North Carter	(21A)
Fg Off E S C Davis	(22B)
Plt Off T N McEvoy	(23B)
Fg Off C J Collingwood	(1920 Dartmouth)
Fg Off D MacFadyen	(20B)
Plt Off G L Worthington	(21B)
Fg Off R L R Atcherley	(22A)
Plt Off H L Patch	(23A)
Fg Off H R D Waghorn	(22A)

8 There is an excellent account in the College Journal 1957 in an article called "Incident from a flight cadet's Log Book."

9 For a fuller account see *Great Mysteries of the Air* by Ralph Barker.

10 H Montgomery Hyde, *Solitary in the Ranks* p. 117. At Cranwell Shaw rode a Brough Superior motorcycle. One presented to him by G B Shaw and his wife was described by T E Shaw as "A very new and large and apolaustic Brough".

11 Oral recording made by author with Mr Cyril Hobbs of Norwood, London SE.

12 H Montgomery Hyde, *Solitary in the Ranks.*

5 The Boys' Wing 1918–1926

1 Letter from Air Council to Treasury on 11 November 1919 in Air 8/10.

2 Some Recollections of Cranwell by Wing Commander S E Townson (1978). Ms in author's possession.

3 In May 1918 the Officer Commanding Boys' Wing was Lieutenant-Colonel A C Barnby, RAF. He was succeeded on 16 December 1918 by Lieutenant Colonel J W L Hunt who handed over to Major R H C Routley on 4 September 1919.
The next two commanding officers in quick succession were Wing Commander G C St P de Dombasle and Wing Commander C F Kilner, DSO. Kilner remained until 1 August 1922 when Wing Commander R J F Barton CBE took over and saw out the unit's remaining time at Cranwell, ie, until it moved to Halton in July/August 1926.

4 Ibid Townson Ms.

5 A valuable list of officers who served with the Boys' Wing is given on pages 34 to 38 of the last issue of the Wing Magazine.

6 Other Ground Units at Cranwell before 1939

1 Operations Record Book (F540)
in Public Record Office. Air 29/718.

2 L/A/A Shirley, TUC—later AVM Sir Thomas Shirley, KBE, CB, DL who became an Engineer Specialist after Cranwell and was AOC and Commandant of the RAF Technical College, Henlow from 1957 to 1959. He was AOC in C Signals Command from 1964 to 1966. The Hyde-Thomson Memorial Prize was founded in 1918 by the generous gift of Mr R D Hyde-Thomson for the promotion of wireless telegraphy and kindred subjects in the RAF.

3 Operations Record Book (F540) in Public Record Office. Air 29/711.

4 Operations Record Book (F540) in Public Record Office. Air 29/833.

7 A Permanent Home for the College

Memorandum 'Future Buildings in the RAF' by CAS on 28 June 1919. Trenchard noted: "Housing of a sufficiently attractive and comfortable nature must be provided as soon as practicable—particularly an Officers Training College, on a plan not inferior to Sandhurst or Dartmouth Colleges, is urgently required."

2 Air Ministry file 72869.

3 Air 2/382.

4 Air 2/396.

5 Air 2/410.

6 Original drawings were prepared in 1920 under the direction of Sir Richard Allison, CBE, FRIBA. Chief Architect, HM Office of Works. Mr J G West was the architect.

7 Air 2/431. Mr J G West, a senior architect, was decorated with the OBE for his work.

8 Files—WORKS 6—364 of 1929/30.

9 The lettering on the foundation stone and on that commemorating the official opening of the College in 1934 was carved by the sculptor, Mr C H Mabey.

10 Lincolnshire—Pevsner and Harris, Penguin Books, 1964.

11 RAF College Journals—Autumn 1930 and March 1950.

8 RAF College Life in the 1930s

1 Information supplied by Mr L S Wigdor of Leas Green, Chislehurst, an old Cholmleian, who won the Aeronautics Prize at Highgate School in 1936. There is an interesting article on the Highgate experiment in the RAF College Journal, Autumn 1929.

2 Thirty officers attended three courses of about six months duration between 1934 and 1936. Among them were Plt Off B R Burnett and Plt Off B Ball on No 1 Course, Plt Off W F Beckwith on No 2 Course and Plt Off A Foord-Kelcey on No 3 Course.

3 Interview with Air Cdre L G Levis (1931-33 C Squadron) in November 1977.

4 In 1929 the Schneider Trophy was won for Great Britain by Flying Officer H R D Waghorn (1922-24 A Squadron) in a Supermarine Rolls Royce S6 Seaplane at an average speed of 328.63 miles per hour.

5 Mr C W E Barnard of Brauncewell, Mr F L Wilson and Mr J G Clarke of Roxholme were presented with an inscribed watch and purse by the Commandant.

6 Flight Lieutenant G E Nicholetts (1921-23 A Squadron) was a last minute replacement for Gayford's original partner, Flight Lieutenant Bett who died in January 1923. Nicholetts had three weeks to grasp the essential details necessary for long distance flying. There is a detailed account of the activities of the RAF Long Range Flight, given as a lecture by Squadron Leader O R Gayford, in the Library of the RAF College.

7 College Journal Vol. XVIII No 1 Spring 1938.

8 Interview with AVM M D Lyne in 1979.

9 See Tables of Commandants and Assistant Commandants.

10 Later Air Chief Marshal Sir Neville Stack, KCB, CVO, CBE, AFC, Commandant 1967-70.

Note on Beagling

Beagles were associated with RAF Cranwell from 1919 until 1936, when the pack was sold.

In 1957 the RAF formed a Beagling Association under the auspices of the RAF Sports Board. Group Captain L G Levis who was commanding No 3 Initial Training School at Cranwell formed the Per Ardua (RAF) Beagles. When he was posted to the Air Ministry between 1954 and 1958, Flight Lieutenant D S Evans and then Flight Lieutenant C H Bidie took over as Master. In 1958 on his return to Cranwell Air Commodore Levis resumed as Master and was sole Master and huntsman for 10 seasons, having in the meantime retired in 1960 from the RAF. In 1968 he was joined in the Mastership and acted as joint Huntsman. The Per Ardua pack ceased to be associated with Cranwell in 1977 and became a civilian pack. Their country lay in the Blankney, Belvoir and South Notts foxhound countries. They also hunted by invitation of the Masters of Foxhounds in the Burton, Southwold, Brocklesby and Cottesmore countries. There were occasional special RAF meets at Cranwell, Wittering, Digby, Coningsby and Nocton Hall.

9 The War Years

1 HQ No 21 Group ORB. Air 25/510.
AVM J E A Baldwin retired from the Royal Air Force at his own request and AVM B E Sutton took over as AOC No 21 Group on 15 August 1939.

2 RAF Cranwell ORB. Air 28/175.

3 RAF College ORB. Air 29/699.

4 No 19 FTS ORB. Air 29/1145.

5 Souvenir Programme included in Air 29/699.

6 No 2 CFS ORB. Air 29/630.

7 No 3 (C) OTU ORB. Air 29/705.

8 Air 28/175.

9 Summary of Training Completed by All Units at Cranwell in 1942.

	Officers	Apprentices	Airmen	Airwomen	Total
No 1 Signals School	480	260	11541	3132	15413
RAF College SFTS	199		360		559
No 3 (C) OTU	142		602		744
No 1 RDF School	263		1746	708	2717
Totals	1084	260	14249	3840	19433

10 Article in RAF News, Oct 24-Nov 6 1979.
11 Air 29/1848.
12 Air 29/725.
13 Air 29/1157.
14 Air 29/711.
15 Air 29/727.
16 Air 29/833.

10 Peacetime and some Difficult Years

1 A G Boycott was an instructor in Imperial and War Studies and wrote a book called *The Elements of Imperial Defence* (1931), often studied by Staff College Q examination candidates in the 1940s.

2 Air Marshal Sir John Slessor who was AMP wrote in a note to the Standing Committee of the Air Council on 25 June 1947: "I feel strongly that there should be a King's Colour for the RAF College Cranwell. Of the 1095 cadets who passed through the College before the war, a total of 555 have been killed in action or by the hazards of flying or are missing and a King's Colour for Cranwell would have a great value as a symbol of loyalty and focus of tradition."

3 RAF Technical College, Henlow. In 1947 the name of the School of Aeronautical Engineering was changed to RAF Technical College, Henlow. It became an air commodore command in August 1949 when Air Commodore D W R Ryley assumed command. In 1952 a scheme of Technical Cadetships was introduced and in 1961 the National Council for Technological Awards recognised the four years and two terms Henlow engineering course for the award of the Diploma in Technology (Engineering). This award was to be reviewed in five years' time, ie the review coincided with the move of the RAF Technical College from Henlow to Cranwell.

4 No 3 Initial Training School ORB Air 29/1831.

5 RAF Cranwell ORB (1946—1950) Air 28/1011; RAF Cranwell ORB (1950) Air 29/1897; RAF College Cranwell ORB (1951 onwards). Held in Air Historical Branch (RAF).

6 RAF College Cranwell. Report of Working Party (Melvin Committee) 1958.

7 Report by HM Inspectors on RAF College Cranwell. Ministry of Education 1958.

8 Air Council 15 (53) on 31 August 1953 and Air Council (53) 45 on 7 September 1953 dealt with this subject. These papers are not yet generally released.

9 Report on the Examination of the Syllabus of the RAF College (Hartley Report) by an Investigating Team appointed by the Air Ministry in 1952

to review the curriculum and organization of the RAF College.
File reference: C46619/51/DTF.
The Air Council endorsed the recommendations of that report in AC (55) 64.

10 The Working Party comprised Air Commodore J D Melvin (D of O) with representatives from DTF, DDTT, DES, DGPS, F8, F4 and DGW. Its Report was presented in April 1958 to the Air Council in a paper AC (58) 34.

11 Antony Constant. For further details see *Who's Who 1978*.

12 Statistics supplied by the Secretariat RAF College Cranwell.

13 AMSO in a report to the Air Council on 6 January 1959.

14 See the ORB (F540) entry for January 1952.

11 The Winds of Change–The 1960s

1 Inspector-General's Report No 518/1954.

2 White Paper on Defence (Cmnd 124 of 1957).

3 The percentages of flight cadets from HMC Schools entering Cranwell between 1958 and 1962 were as follows:

No 78 Entry	63.8
No 79 Entry	51.0
No 80 Entry	50.0
No 81 Entry	48.1
No 82 Entry	34.0
No 83 Entry	31.7
No 84 Entry	39.4
No 85 Entry	22.0
No 86 Entry	23.8
No 87 Entry	21.0

4 AC 19 (60).

5 AC (62) 31.

6 One scheme envisaged both Henlow and Cranwell being kept open with the technical cadets going back to Henlow after their cadet course to complete their engineer officer training.

7 The Newsom Report 1963 had dealt with the question of raising the school leaving age to 16 in 1970/71 and its effect inter alia on the recruitment of boy entrants and apprentices with the introduction of the new Certificate of Secondary Education (CSE) examination.

8 The first RAF Regiment cadet began training at the RAF College on 12 September 1962.

9 All three Services had introduced schemes for University Cadetships in the early 1960s.

10 Report of the Committee on the Implications for the Services of National Developments in Education.

12 The Cranwell–Henlow Merger

1 Trenchard Award Scheme. In August 1960 the Air Council decided to make available from the RAF Central Fund a sum of the order of £2500 a year to be disbursed as Trenchard Memorial Awards to commemorate the unique part played in the history of the Royal Air Force by Marshal of the Royal Air Force Lord Trenchard. The awards are given to support activities of a sporting or adventurous kind that are out of the ordinary or unconventional and are planned on the applicant's own initiative and in his own time.

2 The Selection Board was also located at Cranwell from May to July 1953.

3 Air Force Board. The name of the Air Council became the Air Force Board of the Defence Council in April 1964 when the Ministry of Defence was reorganized under the Secretary of State for Defence.

4 A decision was taken very early on that name tags were to be worn by all officers in uniform at the College. This improved recognition and communication.

5 The cadet strength in November 1966 was:

265 Pilots
38 Navigators
27 Equipment
13 Secretarial
9 Regiment
114 Engineers

6 AFB (66) 21.

7 The decision of the Council for National Academic Awards was given in a letter from the Secretary, Mr Hornby, quoted in file AF/CT3024/64/DFT (RAF).

8 The Lincolnshire Poachers. Over the years the RAF College has had several aerobatic teams but the Poachers formed in 1963. Originally they flew the Jet Provost Mk4 and then the Mk5. They added a distinctive wrap-round flash in light and dark blue to the basic red and white trainer finish of their five aircraft (one was a reserve). The use of the name and especially the use of the Regimental March of the Royal Lincolnshire Regiment was granted to the RAF College by the Colonel of that Regiment in correspondence with the Commandant in 1949. Major-General J A A Griffin wrote:

> I am sure that I speak for all ranks of the Regiment when I say that there is no unit in the Services which we would rather have associated with our Regimental March.
>
> We all very fully appreciate that there would be no Royal Lincolnshire Regiment but for the gallantry, devotion to duty and sacrifice of the officers of the Royal Air Force who fought over Dunkirk and in the

Battle of Britain, many of whom were trained at, and gained experience from, the Royal Air Force College.

9 Training Command. Technical Training Command amalgamated with Flying Training Command to form Training Command on 1 June 1968. It no longer exists under that name since in turn on 13 June 1977 it amalgamated with Support Command to become RAF Support Command.

10 The name of the former RAF Technical Branch was changed to Engineer Branch on 1 October 1966.

11 Senior Specialist Teachers. The RAF Education Branch was permitted to appoint highly qualified officers with predominantly scientific and engineering qualifications to teaching posts not strictly limited by normal establishment criteria. Most of these teachers had Master's degrees or Doctorates. The Scheme began in 1959, the last appointments were made in 1971 and the Specialist Teachers Scheme was abolished in 1972-73, when the majority of those concerned chose to leave the Service.

13 The RAF College 1970–1980

1 It is premature to draw firm conclusions about the success of the Graduate Entry Scheme. Statistics for educational qualifications of officer entrants from civil life in the years 1976-77, 1977-78 and 1978-79 show numbers of graduates are below the planned requirement for the four main executive branches, especially for engineering.

2 Supplementary Statement on Defence Policy 1968 (Cmnd 3701) paras 66 and 67.

3 Hodgkinson Committee. Named after the senior officer who was its Chairman. Later he became Air Chief Marshal Sir Derek Hodgkinson.

4 In June 1978 Prince Charles became the Patron of the Old Cranwellian Association.

5 The student population in July 1972 was as follows:

Graduate Entrants:	56 GD Pilots
	4 Supply
	3 Secretarial
	36 Engineers
Graduate Entrants: (WRAF)	1 Supply
	2 Secretarial
	1 Engineer
Flight Cadets:	29 GD Pilots
	3 Supply
	2 Secretarial
	2 RAF Regiment

6 The strength of the College in November 1975 was as follows:

GE students	212
Other students (RAF)	221
Staff (RAF)	312
Staff (WRAF)	7
Overseas students	77
Exchange officers	12
Warrant officers	20
SNCOs	198
Cpls/ACs	430
WRAF	59
Total	1548

7 Mr J S Robinson was presented with a salver by Air Commodore E D McK Nelson on behalf of all old Cranwellians at a simple ceremony on 20 October 1977. Mr and Mrs Robinson live in retirement in Leasingham.

8 The term 'permanent commission' in this respect implies a commission with full career prospects. Branch commissions, Specialist Aircrew Commissions and Short Service Commissions are excluded.

9 A rather odd reversion to historical pattern has occurred. IOT students spend six weeks in converted barrack blocks, six weeks in the former Junior Cadets' Mess and the final six weeks in College Hall. There are in fact five Officers' Messes at Cranwell now: College Hall, York House, Trenchard Hall and Nos 1 and 2 IOT Messes.

INDEX

Ranks used are as they appear in the text

Printed in England for Her Majesty's Stationery Office by
Hobbs the Printers of Southampton
(33) Dd696772 C30 10/82 G380